Water Security, Conflict and Cooperation in Peri-Urban South Asia

Vishal Narain • Dik Roth
Editors

Water Security, Conflict and Cooperation in Peri-Urban South Asia

Flows across Boundaries

 Springer

Editors
Vishal Narain
Management Development Institute
Gurgaon, India

Dik Roth
Sociology of Development and
Change group
Wageningen University
Wageningen, The Netherlands

ISBN 978-3-030-79037-0   ISBN 978-3-030-79035-6   (eBook)
https://doi.org/10.1007/978-3-030-79035-6

This Springer imprint is published by the registered company Springer Nature Switzerland AG
The registered company address is: Gewerbestrasse 11, 6330 Cham, Switzerland

Preface

This book describes and analyses how urbanization in South Asia changes access to water in peri-urban contexts and how the inhabitants of peri-urban spaces respond to the changes underway. It seeks to address the larger questions of (in-)equity, justice and sustainability that are central to issues of water (in-)security but receive scant attention in mainstream discourses on urbanization, in which this latter process is seen as a necessary and positive step towards development in a way that is conducive to efficiency of resource use, made possible by the economies of scale that cities are able to achieve.

Urbanization has been a key demographic trend in the past and will remain so this century, both globally and in South Asia. Expanding cities tend to be framed as engines of economic growth and development, and as breeding grounds for "smart" and sustainable technologies and lifestyles. Urbanization and the expansion of urban lifestyles can undoubtedly help solving a wide variety of social, economic and environmental problems. For good reasons, growing numbers of citizens all over the world have come to prefer urban life to earlier rural lifestyles and are enjoying the many advantages associated with city life. They may have more economic opportunities, better housing and basic facilities like education and healthcare, water and sewage facilities, and other infrastructure.

There is, however, another side to the story. In a highly unequal world, these urban benefits are not everybody's share, thus many urban inhabitants lack access to the most basic facilities and rights associated with citizenship. Besides this, processes of urban expansion involved often reproduce existing inequalities or create new ones. Urbanization processes are deeply influenced or even largely driven by neo-liberal reform measures and related policy packages. Land speculation, real estate development, growth of outsourcing and information technology sectors, and policies to promote private enterprise have been key drivers of growth in many of them. This investment in capital-driven types of growth is associated with changes in the use and control of land, water and other resources well beyond the city: urban expansion comes through an appropriation and re-allocation of resources away from rural and agrarian activities and lifestyles towards the urban, revealing a bias towards a specific type of urban planning that facilitates the expansion of global private enterprise while at the same time jeopardizing the livelihoods of those who lose their land and have to move and find alternative sources of income.

The contributors to this book explore the peri-urban flipside of the generally positive urbanization narrative. Through its focus on the peri-urban, the book seeks to contribute to the growing body of scholarship on issues of peri-urban water security globally, and particularly in South Asia. "The peri-urban" refers to the spaces changed by urban expansion, basically involving "the coming together and inter-mixing of the urban and the rural, implying the potential for the emergence of wholly new forms of social, economic, and environmental interaction that are no longer accommodated by these received categories" (Leaf 2011, p.528). Focusing on Bangladesh, India and Nepal, the contributions in this book seek to address the following questions:

How does urbanization change access to water in peri-urban contexts? What are the implications of these processes for institutions and practices around water, especially for forms of conflict and cooperation? What kinds of approaches are needed to contribute to the analysis and improvement of peri-urban water security in peri-urban contexts and reconcile competing interests and claims?

The contributions originate in various scientific research programmes and projects. These different origins, as well as the various professional backgrounds and affiliations of contributors, translate into a diverse repertoire of theoretical-conceptual approaches and methods used by the contributors. A wide diversity of themes is addressed in the contributions: questions of urban metabolism and ecological foot-print; gender, rights and access issues; participatory institutional analyses; the institutional analysis and development framework; negotiated approaches; and the formation of multi-stakeholder platforms are some of the themes and approaches that inform scientific and policy discourses on the peri-urban. There is not one framework or analytic lens that is universally applicable to the analysis of peri-urban issues, neither is there a "one-size-fits-all" approach to intervening in peri-urban contexts. As the contributions to this book will show, the peri-urban can be studied and analysed at various scales and levels and through the connections between them.

The book seeks to further the debate on several issues related to water security in peri-urban contexts: what constitutes the peri-urban, including questions of scale and levels; the socially differentiated access to water in peri-urban spaces; appropriate approaches to intervention for improving water access and altering power relations in peri-urban spaces; the implications of the creation of urban infrastructure for peri-urban inhabitants; the diversity of ways in which water serves as a receptacle of urban waste as well as a resource for urban expansion; and the intersection of urbanization and climate change as multiple stressors on peri-urban water resources.

Covering these issues from diverse perspectives, we expect the book to appeal to a range of scholars with various disciplinary backgrounds, groups of professionals working in the worlds of national and international policy, national and international NGOs, activist groups, research and development institutes, and individual readers interested in water security and urbanization, in Bangladesh, India, Nepal and elsewhere.

We hope that the book creates greater awareness of peri-urban water security issues as well as of the need and potential to address it locally, regionally and globally.

Gurgaon, India   Vishal Narain

Wageningen, The Netherlands   Dik Roth

Acknowledgements

Although the contributions in this book draw from several projects across South Asia with which the contributors were engaged, specific ideas for this book took shape during the project meetings of the project on "Climate policy, conflicts and co-operation in peri-urban South Asia: Towards resilient and water secure communities", which was part of the research programme "Conflict and Cooperation in the Management of Climate Change (CoCooN/CCMCC)". This programme was funded by the Dutch Research Council (NWO) of the Netherlands and the Department for International Development (DFID) of the United Kingdom. We thank NWO/DFID for the financial support provided, both for the research and other activities conducted under the project, as well as for the production of this book.

We thank all our contributors for working with us through several versions of their chapters, responding to our queries and requests for clarifications. We also thank the production team at Springer Nature for bearing with us and acceding to our requests for extensions. It is not possible to name the many individuals and organizations who were in some way involved in the research and other activities on which the chapters in this book are based, and provided their time for field visits, interviews, meetings and discussions. Their involvement made it possible for the contributors to investigate their experiences of water (in-)security and their struggles over water.

Gurgaon, India   Vishal Narain

Wageningen, The Netherlands   Dik Roth

Contents

About the Editors

Vishal Narain is Professor of Public Policy and Governance at the Management Development Institute, Gurgaon, India. He holds a PhD from Wageningen University, the Netherlands. His academic interests are the interdisciplinary analysis of public policy processes and institutions, water governance, vulnerability and adaptation to environmental change, as well as peri-urban issues. He is the author of *Public Policy: A view from the South* (Cambridge University Press, 2018) and *Institutions, technology and water control: Water users associations and irrigation management reform in two large-scale systems in India* (Orient Longman, 2003), for which he received the S.R. Sen Prize for the Best Book on agricultural economics and rural development conferred by the Indian Society for Agricultural Economics. He has been a consultant to several organizations such as the Food and Agricultural Organization (FAO) Bangkok; International Water Management Institute (IWMI), Colombo; the Asia Foundation, New Delhi, India; and SaciWATERS, Hyderabad, India. Email: vishalnarain@mdi.ac.in

Dik Roth is a social anthropologist and Associate Professor at the Sociology of Development and Change group of Wageningen University, the Netherlands. He holds an MSc from the University of Amsterdam and a PhD in Social Sciences from Wageningen University. His scientific interests include the anthropology of law and legal pluralism, policy processes, critical development studies, the management and governance of natural resources, resource-related conflicts, and flood risk management policies. His work includes research programmes and activities in Indonesia, India, Nepal, Bangladesh and the Netherlands. He has widely published on various topics in international peer-reviewed scientific journals. He co-edited *Liquid Relations. Contested water rights and legal complexity* (Rutgers University Press, 2005) and *Controlling the water. Matching technology and institutions in irrigation management in India and Nepal* (Oxford University Press, 2013). Email: dik.roth@wur.nl / droth@xs4all.nl

About the Contributors

Madhureema Auddy works as a programme coordinator at Gramin Vikas Vigyan Samiti (GRAVIS) in Jodhpur, India. She holds an MA in Development from Azim Premji University. Her interests lie in exploring issues on water and waste management, climate action and sustainability from a systems approach. Email: madhureema@gravis.org.in

Diganta Das is an urban studies scholar and Associate Professor at Humanities and Social Studies Education, Nanyang Technological University. He received his MSc from Jawaharlal Nehru University (JNU), India and M.Phil. from the Indian Institute of Technology (IIT), Bombay. He received his PhD from the National University of Singapore (NUS). His research interests focus on relations between the production of smart cities, high-tech space-making and issues of human agency in urban Asia, public policy and mobility; changing dynamics of urban waterscapes and issues of liveability and sustainable urban development. He is currently involved in several research projects. In two of his ongoing research projects, he examines urban waterscapes and how changing urban dynamics and global aspirations affect the waterscapes of Asian cities. The second project intends to provide detailed genealogies of smart cities through in-depth empirical research on the contexts of India and South Africa. Email: diganta.das@nie.edu.sg

C. S. Dechamma Dechamma C.S works at the Centre for Climate Change and Sustainability at Azim Premji University. She holds an MA in Development with a specialization in sustainability from Azim Premji University. Her interests lie in the area of urban commons, specifically in the conservation and management of lakes in Bengaluru. Email: dechamma.cs15@apu.edu.in

Sharlene L. Gomes is currently a post-doctoral researcher in the Faculty of Technology, Policy and Management of Delft University of Technology, Delft, the Netherlands. She holds a PhD from Delft University of Technology, and an MSc in Water Science, Policy, and Management from the University of Oxford. Her research focuses on the institutional and governance context of peri-urban areas, with an emphasis on the water sector. She is interested in trans-disciplinary methods to engage stakeholders in policy analysis and policy planning activities in urbanization contexts. She has worked in several cities in India and Bangladesh on these topics as part of the *Shifting Grounds* and *H2O-T2S in urban fringe areas* projects. Email: S.L.Gomes@tudelft.nl

Sadika Haque is an agricultural economist and professor at Bangladesh Agricultural University. Along with teaching, she has been engaged with several research projects, focusing on economics of agricultural production, gender, poverty, migration, nutrition, women's empowerment and natural resource management. In her PhD thesis, she showed how local users lose their rights due to the entrance of outsiders, if inequality increases in common resource-dependent rural areas. Sadika is also an activist in her fields of research, and has published a number of articles on migrant ready-made garment industry workers, tea garden workers, female labour in agriculture, microcredit borrowers, common resource users, poor labourers working in urban areas, female farmers' rights, food sovereignty etc. She has been working with USAID, UKAID, and different CGIAR research organizations. Along with other researchers, she has also written several policy briefs based on the issues she was engaged with through her research. Email: sadikahaque@gmail.com

A. T. M. Zakir Hossain is the founder and executive director of Jagrata Juba Shangha, a leading non-profit organization in coastal Bangladesh, based in Khulna. He is a development practitioner and socio-environmental expert with more than 35 years of experience in the development sector. He received his MSc degree in Commerce from Rajshahi University, Bangladesh, and a certificate course on an International Project Managers' programme from Globalverkstan, Gothenburg, Sweden. He has expertise in socio-environmental and capacity-building research, and in project design and implementation, human resource management, liaison and coordination. Mr. Hossain possesses extensive partner management experience gained through different partnership-based projects. Climate change, environmental change, disaster management, food security, gender, child protection and governance are his major areas of engagement. Mr. Hossain co-authored two scholarly publications related to capacity-building for water management in Bangladesh. Email: atmzakir@gmail.com

Sheikh Nazmul Huda is project coordinator of the MOHORA project for Jagrata Juba Shangha based in Khulna, Bangladesh. He is a development worker with 20 years of experience in this field. He graduated in Commerce from the National University of Bangladesh. He has expertise in socio-environmental and capacity-building research, and demonstrated skills in project design and implementation, liaison and coordination. Mr. Huda has experience in dealing with climate change, environmental change, disaster management, food security, gender, child protection and governance issues. He has co-authored scholarly publications related to capacity-building for participatory water management in Bangladesh. Email: nazmuljjs@gmail.com

Deepa Joshi is the Gender, Youth and Inclusion Lead at the CGIAR Research Programme on Water, Land and Ecosystems (WLE) of the International Water Management Institute (IWMI). A feminist political ecologist by training, her research has analysed shifts in environmental policies and how these contextually restructure complex intersections of gender, poverty, class, ethnicity and identity. Her interests lie in connecting gender and environmental discourse to local capacity-building initiatives and advocating for policy-relevant change across developmental institutions. She has worked primarily in South Asia, and to a lesser extent in Southeast Asia, Sub-Saharan Africa and Latin America. Her published research presents ethnographic analyses of how inequality is reiterated and experienced across institutions and processes of policymaking, both in policies and in implementing institutions at scale. Email: deepa.joshi@cgiar.org

M. Shah Alam Khan is professor at the Institute of Water and Flood Management of Bangladesh University of Engineering and Technology. He is a coastal and urban water management expert with more than twenty years of experience in research and teaching. His advanced training and collaborative work with multidisciplinary partners in several countries led him to develop a broader interdisciplinary understanding of coastal and urban water challenges. This placed him among the leading scientists with the capacity to integrate methods and approaches from various disciplines including water engineering, natural sciences, social sciences, spatial planning and landscape architecture. In his research, Professor Khan explores critical issues related to water-, nature- and climate-sensitive infrastructure, urban flooding, sustainable urban and coastal development, adaptive water management, decision-making under uncertainty, community resilience, and inclusive planning and governance. Email: msakhan.buet@gmail.com

Nathaniel Dylan Lim is a graduate from the National Institute of Education, Nanyang Technological University, Singapore where he received his BA (Education) in Geography with highest distinction. His research interests include the everyday politics of water resource management, geographical inquiry and the geographies of education. His BA thesis on the everyday politics of water (in)security has been awarded the highest performing undergraduate coursework in the Social Sciences: Anthropology & Cultural Studies category in the Global Undergraduate Awards 2020. Email: nathanieldylan@gmail.com

Pratik Mishra is a PhD student in Geography at King's College London. His research interests span largely around urban metabolism and political ecology, and further into labour geography, infrastructure, migration, and climate change. These interests have been variously covered within his pre-doctoral work on urban canal systems, and his doctoral fieldwork on brick kilns. His recent papers examine the political ecology of the operation of kiln clusters in Delhi, and on migrant circulations of brick kiln workers between the kiln and the city. His articles have appeared in *Economic and Political Weekly, Urbanisation*, and the *South Asian Multidisciplinary Academic Journal*. Email: pratik.mishra@kcl.ac.uk

Seema Mundoli is a senior lecturer at Azim Premji University. She began her career in Human Resources in the corporate sector, and later did her MA in Development. Her work looks at social and ecological interactions around urban ecosystems in Indian cities. She has co-edited the *State of the environment, Andaman and Nicobar Islands* (Universities Press India 2005) and co-authored *Cities and canopies: Trees in Indian cities* (Penguin India, 2016, with Harini Nagendra). Email: seema.mundoli@apu.edu.in

Harini Nagendra is a Professor of Sustainability at Azim Premji University. She holds a PhD in Ecology from the Indian Institute of Science. Her work focuses on human-nature interactions in cities and forests in the global South. Her books include *Nature in the city: Bengaluru in the past, present and future* (Oxford University Press India 2016) and *Cities and canopies: Trees in Indian cities* (Penguin India 2016; with Seema Mundoli). Email: harini.nagendra@apu.edu.in

Kamrun Nahar is pursuing her MSc at the department of Agribusiness and Agricultural Economics, University of Manitoba, Canada. She completed her BSc and MSc in Agricultural Economics from Bangladesh Agricultural University. During her MSc in Bangladesh Agricultural University she completed a gender research in the beginning of her career life where she collected information by living with her rescarch participants (ready-made garment workers). Her major areas of research interest are production economics, gender and women's empowerment. Email: knahar.bau13@gmail.com

Mohammad Rezaur Rahman is a civil engineer by profession and professor at the Institute of Water and Flood Management of Bangladesh University of Engineering and Technology (IWFM, BUET). He specializes on environmental aspects of water sector planning and management. The post-graduate courses that he has taught include Socio-Economic Assessment, Water and Ecosystem, and Climate Change Risk Management. He has been involved in a number of international collaborative interdisciplinary research projects. His current research projects include REACH: Improving Water Security for the Poor, led by the University of Oxford. He has supervised a number of post-graduate theses on water- and environment-related issues, especially in the Khulna region. He has participated in the preparation of several national plans and strategies, including the National Water Management Plan, the National Sustainable Development Strategy and the Bangladesh Climate Change Strategy and Action Plan. Email: rezaur@iwfm. buet.ac.bd

Abhiri Sanfui is a research associate at Azim Premji University. She has done her Masters in English Literature from Jadavpur University. She has been a part of the Youth Community Reporters' project, which is a collaboration between Jadavpur University and UNICEF. Her area of interest is Dalit Literature and Ecofeminism. Email: avisanfui@gmail.com

Anushiya Shrestha has a PhD from Wageningen University, The Netherlands. She is a researcher at the South Asia Institute of Advanced Studies (SIAS), a recognized research organization in Nepal. Her research interests include the policies and practices around changing resource use and management, focusing especially on peri-urban water issues. She has published several articles on peri-urban water security in Nepalese and international peer-reviewed journals. Since 2020, she is involved in a research project that focuses on understanding and enhancing the political capabilities of marginalized groups in an urban context. Email: anushiya.shr18@gmail.com

Jasber Singh is Associate Professor at the Centre for Agroecology, Water and Resilience at Coventry University. He has several years of experience in designing, delivering and evaluating participatory action research projects on social and environmental justice at the local, national and international levels. His interests are creative participatory action research approaches, such as participatory theatre and photography – which he applies on issues of environmental justice, decoloniality, anti-racism, youth participation, migrant and refugee rights, the politics of difference, food sovereignty, social movements, right to foods, agrarian distress and pedagogy. He has worked with local NGOs in India to document how racism undermines food, gender and land rights, and he has advocated for change at the local level and with policymakers working with the supreme court of India. Email: ac5866@coventry.ac.uk

Shahinur Tania is working as Monitoring Officer at the LDDP (Livestock and Diary Development) project and studied at the Department of Agricultural Economics of Bangladesh Agricultural University. In the research reported in this book, along with Kamrun Nahar she stayed in the study area for six months to observe the lives of ready-made garment (RMG) industry workers. Basically, she considered the reasons for migration of the RMG workers and the actual changes happening through this kind of migration in their lives. Currently she is further developing her research skills in further research work with Professor Dr Sadika Haque. She is very much interested in working on the social aspects of migration, gender, women's empowerment. Email address: shahinurtania.bau@gmail.com

Nusrat Jahan Tarin is an MSc student at the Institute of Water and Flood Management of the Bangladesh University of Engineering and Technology (IWFM, BUET). She has an environmental science background from the Soil, Water and Environment Department (SWED) of the University of Dhaka. Having affinity with scientific research and being concerned about issues regarding the environment, she preferred an interdisciplinary study in the WRD programme at IWFM, BUET. Through her research work she became involved in several research projects on renewable energy, ecosystem conservation, climate change, water security issues in peri-urban setting, conflict resolution, resource management, socio-economic progress, etc. She gained experience in delta management while working in a capacity-building project directly related to BDP 2100. With her research interest on WASH, public health and gender studies, Tarin worked in a team to develop better understanding of a socio-technical research approach. She co-authored publications on the environment, social science and development. Email: nusrat.tarin08@gmail.com

Sumit Vij holds a PhD from Wageningen University, the Netherlands. He currently works as a postdoctoral researcher at the Faculty of Sciences of Free

University Amsterdam, and with the Public Administration and Policy Group of Wageningen University and Research. His research interests are climate governance, power and politics, transboundary water politics and peri-urban development. Between 2007 and 2009, he worked on the National Dairy Plan at the National Dairy Development Board of the Indian Ministry of Agriculture. From 2013 to 2015 Sumit worked as a researcher for various climate change adaptation and water security projects in India, Nepal and Bangladesh. He has published in various international journals such as *Environmental Science and Policy, Climate Policy, Environment Development and Sustainability, International Journal for Water Resources Development, Land Use Policy, Water, Water International, Water Policy, Action Research, WIREs Water* and *Geoforum.* Email: sumit.vij@wur.nl

Tina Wallace is a feminist, gender advisor, long-time development researcher, teacher and practitioner, consultant. She has experience in teaching, research and practice in the field of development and has widely published in these areas of work. She currently runs seminars and workshops with the Development Studies Association and International Gender Studies at Oxford University. She has worked with a range of INGOs in the United Kingdom, Africa, Middle East and Asia on issues around learning, strategic thinking, addressing gender inequalities and women's rights. She is on the advisory body of a number of foundations in grant making and editorial roles, as well as in undertaking new research (currently with IIEP, UNESCO). Email: tinacwallace@icloud.com

Saroj Yakami has an MSc in International Land and Water Management (MIL) from Wageningen University, the Netherlands. Currently he is project manager with MetaMeta in Nepal, and leading work related to water resource management, agriculture, and climate change. He has worked on issues of water management, water security, disability and agriculture, capacity-building, and is experienced in working with communities. His areas of interest are water management, agriculture and climate change. He is currently working on a "green roads for water management" project in Nepal. Email: syakami@metameta.nl.; yakami.saroj@gmail.com

Chapter 1
Introduction: Peri-Urban Water Security in South Asia

Vishal Narain and Dik Roth

1.1 Setting the Scene

The world is rapidly urbanizing. With around 55 per cent of the world's 7.63 billion people living in urban areas (United Nations, 2019) we are facing conditions of "planetary urbanism" (Friedmann, 2016) and "planetary urbanization" (Brenner & Schmid, 2012; Swyngedouw & Kaika, 2014). The global urban population is expected to grow by 2.5 billion between 2018 and 2050, with nearly 90 per cent of this increase concentrated in Asia and Africa. An estimated 68 per cent of the world's population will reside in urban areas by 2050. Almost half of the urban population currently lives in urban settlements of less than 500,000 inhabitants, rather than in the relatively few mega-cities of the world (United Nations, 2019).

This trend is expected to continue: much future urban growth will probably take place in a large number of smaller cities with a population of one million or less in Asia and Africa (United Nations, 2015, 2019; see also Satterthwaite, 2006). In the prospects for 2018–2030 for these relatively less urbanized regions, the number of cities with 500,000 or more inhabitants is expected to grow by 57 per cent in Africa and by 23 per cent in Asia (United Nations, 2019, p. 11). The same report estimates that "all the expected world population growth during 2018-2050 will be in urban areas": while the urban population is expected to rise from 4.2 billion to 6.7 billion,

Author sequences for Vishal Narain and Dik Roth are alphabetical: both authors have equally contributed to the writing of this chapter and to the editing process of the book.

V. Narain (✉)
Management Development Institute, Gurgaon, India
e-mail: vishalnarain@mdi.ac.in

D. Roth
Sociology of Development and Change group, Wageningen University,
Wageningen, The Netherlands
e-mail: dik.roth@wur.nl

© The Author(s) 2022
V. Narain, D. Roth (eds.), *Water Security, Conflict and Cooperation in Peri-Urban South Asia*, https://doi.org/10.1007/978-3-030-79035-6_1

the total world population is projected to grow from 7.6 billion in 2018 to 9.8 billion in 2050. Three sources mainly account for this urban growth: natural increase, rural-urban migration, and the expansion of cities, leading to annexation and transformation of rural areas into urban settlements (United Nations, 2019; see Leaf, 2016).

In this book we specifically engage with this last-mentioned dimension of urbanization: the ongoing expansion of cities into their rural surroundings, and the multiple water security problems resulting from these processes. Our focus is on those spaces that are transformed by urban expansion, often called "peri-urban" (Friedmann, 2016; Leaf, 2011; United Nations, 2015). This term refers to "the coming together and intermixing of the urban and the rural, implying the potential for the emergence of wholly new forms of social, economic, and environmental interaction that are no longer accommodated by these received categories" (Leaf, 2011, p.528). As a fluid resource, water is symbolic of the wider socio-ecological flows of urbanization that deeply influence the peri-urban. Taking an "underall" view of changing peri-urban water security, the book explores the *flows across boundaries* that are crucial for understanding the changing water uses, rights and controls, as well as in- and exclusions that determine water security in peri-urban spaces.

The regional focus of this book is South Asia, where urbanization has been, and will remain, a key demographic trend in the decades to come. Its urban population has grown by 130 million between 2001 and 2011, and is expected to grow by another 250 million in the next 15 years. Six of the world's mega-cities—Bangalore, Delhi, Dhaka, Karachi, Kolkata and Mumbai— are located in this region, with others (Ahmedabad, Chennai, Hyderabad, Lahore) following suit (Ellis & Roberts, 2016). The contributing authors explore water security in the peri-urban spaces of cities in three countries: Bangladesh, India and Nepal. In South Asia and elsewhere, the growth of cities entails radical changes in the control and use of nature's "resources" like land and water. The contributors to this book describe and analyse how urbanization changes access to and control over water in various peri-urban contexts, and how the inhabitants of peri-urban spaces experience and respond to these changes. More specifically, they seek to address the following questions:

How does urbanization change access to water and water security in peri-urban contexts? What are the implications of these processes for institutions and practices around water, especially for forms of conflict and cooperation? What kinds of approaches are needed to contribute to the analysis and improvement of peri-urban water access in peri-urban contexts and reconcile competing interests and claims?

The book adds to a growing body of scholarship on the peri-urban and, more specifically, on peri-urban water security in South-Asia (for a review, see Narain & Prakash, 2016). Although scientific and policy interest in the peri-urban, its emergent and often messy character, and its problematic linkages to urbanization have considerably increased in the last decades, on the whole such attention is still relatively marginal. Despite a growing body of work on the peri-urban by urban(ization) scholars, geographers and urban political ecologists,[1] thinking in terms of an urban-rural dichotomy is still quite prominent, especially in the policy world. Such neat, often territorially defined and administratively fixed, categories provide an illusion of orderliness and manageability that continues to be reproduced in policies and intervention-focused research (see Arabindoo, 2009). More importantly, according to Angelo and Wachsmuth (2020) the last few decades have seen an "urban turn": often cities are no longer framed as part of the world's sustainability and development problems, but as a contribution to solving these problems (Angelo & Wachsmuth, 2020), mainly through the large-scale application of "smart" and "sustainable" technologies that reduce the ecological footprint. This framing of urbanization and the urban condition is most prominently expressed in ecological modernization thinking and practices and in the "sustainable" urban agendas that have been developed on its basis (Angelo & Wachsmuth, 2020; see Keil, 2020; see also below).

Although we do not deny that urbanization and the expansion of urban lifestyles can help solving a wide variety of social, economic and environmental problems, we argue that more in-depth attention to the peri-urban dynamics of urbanization provides crucial insights into the peri-urban flipside of this positive urbanization narrative. As the contributions to this book will show, peri-urban populations carry the burden of the expansion of cities in many ways. They experience and have to adapt their livelihoods to radical —often speculative and capital-driven (see Shatkin, 2016, 2019; Simon, 2008)— changes in land use, land prices and land control, and growing densities of building and infrastructure catering primarily to private investors. These changes also deeply influence peri-urban water security: while peri-urban water flows are increasingly controlled and used to provide urban dwellers and other urban users with freshwater, growing problems of pollution, excessive groundwater withdrawal and surface water depletion, and solid and effluent waste disposal threaten peri-urban land and water. To make things worse, public water provision systems tend to bypass peri-urban areas, leaving peri-urban dwellers dependent on their own alternative needs-driven access strategies and practices based on traditional water sources, the use of new technologies, privatized provision etc. (see e.g. Allen et al., 2006; Shrestha, 2019). The chapters of this book explore such dimensions of peri-urban water insecurity, including growing competition over groundwater, growing stresses over lakes and wetlands and the socio-technical mediation of water insecurity along freshwater canals in a "no longer rural, not yet urban" setting, and wastewater canals that run across rural and urban areas.

This chapter first introduces the various perspectives, themes and cases presented in the book chapters. It then discusses urbanization and the peri-urban more specifically, introducing two contrasting views — ecological modernization and political ecology — and introduces the concept of water security. Referring to the examples

---

[1] For a small selection, see e.g. Allen, 2003; Allen et al., 2006; Leaf, 2011, 2016; Satterthwaite, 2016; Shatkin, 2016; Simon, 2008; Tacoli, 1998; for urban political ecology, see e.g. Kaika, 2017; Swyngedouw & Heynen, 2003; Swyngedouw, 2009; see also below). For South Asia, see e.g. Narain & Prakash, 2016.

from the book, the chapter then gives an overview of some of its key themes: the role of material infrastructure; property transformations and the declining commons; socially differentiated access to water; intervening in the peri-urban; and the role of conflict and cooperation.

1.2 Peri-Urban Cases and Approaches

1.2.1 Selection of Peri-Urban Cases

Each of the countries featuring in this book has its own specific population and urbanization histories and characteristics. Bangladesh has a current total population of 165 million, around 40 per cent of which is urban. While the country's rural population is expected to decrease by 20 per cent (around 21 million people) between 2018 and 2050, the country will contribute more than 50 million to urban growth in the same period (United Nations, 2019). India currently has a 1.35 billion population. Its urbanization level, below 20 per cent in 1950, has almost doubled to 34 per cent (461 million urban inhabitants) by 2018. However, with 893 million, India still has the world's largest rural population. While India has five megacities, 55 per cent of its urban population lives in cities with less than one million inhabitants. In the 2018–2050 period, India is estimated to contribute another 416 million urban dwellers and thus almost double its urban population size again. In the same period, its rural population will decrease by around 111 million (United Nations, 2019). Nepal has a current population of around 29 million. With a 19.7 per cent urban population in 2018, it is also among the least urbanized countries in the world.[2] However, the country's urbanization level is expected to rise to 30 per cent by 2050. This growth will especially take place in Kathmandu Valley, in which the country's capital Kathmandu is located (see Muzzini & Aparicio, 2013). Its rate of urbanization (2.9% in 1990–2018; ranking fifth in the world) will decrease to 2.0% in the period 2018–2050 – ranking second in the world (United Nations, 2019).

The choice of urbanization and peri-urban cases in these countries is partly based on specific peri-urban water security issues that drew the attention of the contributors to this book, and partly on more pragmatic considerations such as the opportunity to build on earlier research projects, the existence of academic and NGO networks to cooperate with, and community engagements that made forms of action research possible. The resulting chapters cover research on peri-urban Dhaka and Khulna in Bangladesh, Bengaluru, Gurugram, Hyderabad, Kolkata and Pune in India, and Kathmandu (Valley) in Nepal. Three chapters present peri-urban case studies from India (Mundoli et al., Chap. 2; Lim and Das, Chap. 5; Mishra and Vij, Chap. 6), two are based on research in Bangladesh (Joshi et al., Chap. 4; Shah Alam

---

[2] It shares this characteristic with Sri Lanka, also in South Asia. Nepal's level of urbanization was 2.7% in 1960 and grew to 8.9% in 1990).

Fig. 1.1 Queuing up for water in peri-urban Hyderabad. (Photo Dik Roth)

et al., Chap. 7), while one chapter refers to cases from both Bangladesh and India (Gomes, Chap. 8), and another is on Nepal (Shrestha et al., Chap. 3). The chapters deal with various water security problems: water-waste linkages around lakes and wetlands (Mundoli et al.,); changing irrigation infrastructure and water uses, and the emergence of alternative water sources for irrigation (Shrestha et al.,; Mishra and Vij), water, agriculture and climate change (Mishra and Vij); the marginalized position of female workers and lack of access to water and sanitation facilities in the ready-made garment industry (Joshi et al.,), everyday experiences of peri-urban water insecurity in a context of urbanization-driven depletion and privatization (Lim and Das; Gomes); and participatory approaches to solving conflicts around contested water control infrastructure and ways towards solving them (Shah Alam et al.,) (see Fig. 1.1).

1.2.2 Various Engagements, Themes and Perspectives

The contributors to this volume represent a cross-section of academics, researchers, development practitioners and water professionals in Asia and Europe, including both senior and early-career researchers. Their research activities originate from various research projects with different academic and societal objectives that have co-determined the issues, questions and forms of engagement of project contributors.[3] The different disciplinary orientations, professional backgrounds and societal engagements of contributors further mean that there is not one single conceptual or theoretical framework, research method or type of data that informs all approaches to the peri-urban as a field of research presented in this book.

Thus the chapters reflect the various ways in which academics and other professionals in Bangladesh, India and Nepali are engaging with these processes, how to research and analyse them, and how to contribute to improving the conditions of those who are at the losing end in terms of their water security. Some chapters have a mainly descriptive and critical analytical focus on changes underway in water access and water security (Mundoli et al.,; Shrestha et al.,; Joshi et al.,; Lim and Das; Mishra and Vij). Others engage with the development, application and improvement of approaches to intervention in peri-urban spaces, the need for which is rapidly growing (Shah Alam et al.,; Gomes). These latter chapters show explicit engagement with policy issues through action research and participatory approaches.

Conceptually and theoretically, the chapters are influenced by approaches like political economy and political ecology, legal anthropology, commons studies, participatory institutional analysis, development policy analysis, and negotiated and multi-stakeholder approaches. Overall, the contributions engage with issues of water security and water rights, vulnerability and resilience, gender and other mechanisms of social differentiation, equity and justice. The diversity in the units of analysis, scales and scalar relationships researched in the various chapters suggests that the peri-urban needs not necessarily be understood exclusively as a geographical area demarcated at the periphery of cities, as is still commonly assumed. Peri-urban issues can be examined at various scales and levels, the complex interlinkages between them being crucial. The chapter by Pratik Mishra and Sumit Vij, for instance, takes as its unit of analysis a zone of three canals that run parallel to each other, cutting across rural and urban areas. Its unit of analysis is not an area at the periphery of a city, but rather a water supply infrastructure that straddles the rural-urban divide. In contrast, Mundoli et al. focus on a lake and a wetland as sites for investigating peri-urban dynamics around water, particularly in how a peri-urban conceptual lens helps us analyse the urban metabolism and ecological footprint of cities. Joshi et al. focus on female ready-made garment workers to highlight the socially differentiated access to water in peri-urban contexts and their daily struggles to access water, which adds to the already high work burdens at home.

---

[3] Several chapters originate from the project "Climate policy, conflicts and cooperation in peri-urban South Asia: towards resilient and water secure communities", which was part of the research program Conflict and Cooperation in the Management of Climate Change (CoCooN/CCMCC), funded by the Dutch Research Council (NWO), the Netherlands, and the Department for International Development (DFID) of the United Kingdom.

1.3 Urbanization, the Peri-Urban and Water Security

1.3.1 The Urban and the Peri-Urban

The peri-urban should be understood in relation to the processes of "urbanization of nature" and the socio-environmental changes that are at its core (Swyngedouw, 2009). Swyngedouw and Kaika (2014, p.465) distinguish three key perspectives on "the urban environmental question": urban sustainability, urban environmental justice and urban political ecology. While fully acknowledging the relevance of environmental justice approaches and the role of social movements, we focus on the other approaches here, as these are most relevant for discussing the peri-urban more specifically.

In the last few decades, urban conditions and lifestyles have made a remarkable come-back in environmental and development policies. From major problem sites, hotbeds of widespread poverty, expanding slums and environmental degradation, cities have become part of the perceived solution to major world problems, primarily environmental and developmental. Cities and urbanization processes are widely framed nowadays as basically beneficial and sustainable, as long as the right ("smart") techno-managerial arrangements are in place (see Cook & Swyngedouw, 2012; Swyngedouw & Kaika, 2014). The United Nations report cited above, for instance, states that:

> Urbanization has generally been a positive force for economic growth, poverty reduction and human development. Cities are places where entrepreneurship and technological innovation can thrive, thanks to a diverse and well-educated labour force and a high concentration of businesses. Urban areas also serve as hubs for development, where the proximity of commerce, government and transportation provide the infrastructure necessary for sharing knowledge and information. (2019, pp. 1-2).

If urbanization is acknowledged to be a problem at all by threatening "sustainability", it is regarded as a basically technical and managerial one: "Unplanned or inadequately managed urban expansion, in combination with unsustainable production and consumption patterns and a lack of capacity of public institutions to manage urbanization, can impair sustainability due to urban sprawl, pollution and environmental degradation." (2019, p.1). This framing of urbanization and the urban condition is most prominently expressed in the developmental claims, assumptions and approaches of ecological modernization thinking, in which it is argued that "human development is becoming delinked from the processes that cause environmental degradation" (Clement, 2010, p. 141; Keil, 2020; see also Kallis & Bliss, 2019).

The basic ideas of ecological modernization have increasingly influenced urban and urbanization scholars in developing agendas for "urban sustainability" (Clement, 2010; Cook & Swyngedouw, 2012). Notions like "smart growth", "smart cities", "green urban development" and "sustainable cities" are more popular than ever before, leading Angelo and Wachsmuth (2020, p.2202; see also Keil 2020) to the conclusion that "sustainable urbanism has become a new policy common sense". Kaika (2017) cites UN-Habitat's (2010) report *Cities for All: Bridging the Urban*

Divide which, in contrast to earlier reports, describes urbanisation as a "'positive force for transformation' " in the Global South and noting that 'too many countries have adopted an ambivalent or hostile attitude to the urbanisation process, with negative consequences' (2010: 26)".[4] Like the ecological modernization, on which it is based, "sustainable cities" thinking is explicitly market- and growth-based, focusing on techno-managerial and related institutional and governance principles, paying little attention for issues of inequality, conflict and justice (see Kaika, 2017; Swyngedouw & Kaika, 2014). Cugurullo (2016, p.2421), discussing Abu Dhabi's flagship "eco-city" Masdar City, characterizes ecological modernization as "one of the most international manifestations of the ideology of sustainability", in which "the city is treated as a commodity and its development is dictated by the logic of the market." (2016, p.2430).

Similarly, as an example of this kind of urban development discourse and the practices related to it, Kaika (2017; referring to work by Datta (2015), mentions India's "smart cities" programme as a form of smart city promotion with "highly questionable socio-environmental outcomes, becoming at best a form of 'entrepreneurial urbanization'" (Kaika, 2017, p.91). One of our cases in this book (Misra and Vij) concerns Gurugram, the "model" smart city for India's smart cities project.[5] Gurugram is an example of the impacts of a type of urbanization planned as a regional industrial and commercial centre in neo-liberalized India of the 1990s. It caters to those who are on the winning side of India's economic development. Planning was minimal and building largely run by real-estate developers of companies and housing projects, targeting companies and well-to-do higher middle class seekers of housing. Renamed Gurugram from its earlier, more popular name "Gurgaon", it is now propagated as "Cyber city of Haryana" (the state in which it is located). Some of the downsides of this neoliberal success story, its peri-urban agricultural and water use practices, are discussed in Chap. 6.[6]

Approaches to urbanization developed in urban political ecology stand in sharp contrast to the ecological modernization approach discussed above. Basic to urban political ecology is its rejection of an ontological divide between nature and society, approaching them instead as mutually constituted "socionatures" or "socio-natural assemblages" (Swyngedouw & Heynen, 2003). Urbanization is analysed as a "process of geographically arranged socio-environmental metabolisms that fuse the social with the physical" (Swyngedouw & Kaika, 2014, p.465). According to the

---

[4] Bringing in more buzzwords of the day, this line is continued in United Nations SDG goal 11: make cities inclusive, safe, resilient and sustainable, and in the New Urban Agenda of Habitat III; see https://www.un.org/sustainabledevelopment/cities/; http://habitat3.org/the-new-urban-agenda. For criticism see Satterthwaite (2016) and Kaika (2017).

[5] See:https://www.indiatoday.in/mail-today/story/gurgaon-smart-city-pm-narendra-modi-nda-250840-2015-04-30

[6] See Gurugram's website on: https://gurugram.gov.in. In contrast, the work by the French photographer Arthur Crestani painfully expresses this type of development by portraying those for whom the new city has no place; see http://arthurcrestani.com/bad-city-dreams-7/. For inequalities in the city's sewage infrastructure development, see Gururani (2017).

same authors, "the urban process has to be theorized, understood and managed as a socio-natural process that goes beyond the technical-managerial mediation of urban socio-ecological relations" (p.466). This is a crucial dimension of urban political ecology, as it brings within view the processes of mobilization, reallocation and commodification of nature across scales and boundaries and the resource flows resulting from them, which form the core of the processes of urban metabolism, and hence also of (peri-)urbanization (see Swyngedouw & Kaika, 2014). The metabolization of nature is not a neutral process: it both reflects existing power relations and inequalities in a capitalist society and produces new ones, an ongoing multiscalar process that creates new benefits for some and burdens for others, in- and exclusions, environmental injustices, as well as the socio-political contestations that are part of these processes (Swyngedouw & Kaika, 2014).

Although sympathetic and theoretically close to urban political ecology, Angelo and Wachsmuth (2020, p.24) are also critical of how the research agenda of urban political ecology has developed. They argue that even critical studies based on urban political ecology, in which processes of transformation of nature have a central place, suffer from "methodological cityism", in which "the city has remained the privileged lens for studying contemporary processes of urban transformation that are not limited to the city". Even though authors like Heynen and Swyngedouw stress that there "is no longer an outside or limit to the city, and the urban process harbours social and ecological processes that are embedded in dense and multilayered networks of local, regional, national and global connections" (2003, p.899), Angelo and Wachsmuth (2014, p.24) state that "actually-existing UPE is mainly a research program into the politics of nature within cities". Webster (2011) also stresses the need to shift the balance from a city perspective towards a rural and peri-urban counter-perspective (for urbanization and rural transformation in China, see Muldavin, 2015).

Water security is an emerging and much debated paradigm (for an overview of how the paradigm evolved, see Cook & Bakker, 2012). It has a wide variety of, often contradictory, connotations and is used by proponents of different disciplinary orientations and backgrounds. However, it remains a relevant conceptual lens to study peri-urban processes, as it gives insight into the processes of resource reallocation consequent upon urban expansion, as the contributions in this book demonstrate. While recent years have seen a rising interest in issues of peri-urban water security in South Asia (as examples, see Narain & Prakash, 2016; Roth et al., 2018a, b), the contributions in this book take the analyses further to explore the implications of increasing peri-urban water insecurity for institutions around water and emerging forms of conflicts and co-operation.

Interdisciplinary peri-urban research that explores specific situated dimensions of the socio-ecological flows associated with the urbanization of nature (such as flows of water in this book) can be an important addition to the current, primarily urban-focused research and scientific literature (see Bartels et al., 2020). A research focus from such an interdisciplinary perspective on the constitution of the peri-urban through these processes, how they are locally experienced, perceived and acted upon, including the power differences, disjunctures, inequalities and exclusions that are emerging and are being reproduced or transformed in the shaping of peri-urban spaces, can contribute to a better understanding of these processes, while avoiding methodological "cityism". In contrast to disciplinary urban planning or engineering approaches, an interdisciplinary perspective could encompass the rich insights from political economy, political ecology, sociology, social anthropology, human geography, social studies of science, actor-network theory and many more. Much peri-urban focused work has already been done from such a perspective on topics like water security and water rights (for South Asia see e.g. Karpouzoglou & Zimmer, 2016; Karpouzoglou et al., 2018; Mehta et al., 2014; Narain & Prakash, 2016). Such research could, however, be more explicitly integrated with critical urban research agendas, to put into perspective the "sustainable cities" narrative by explicitly showing its peri-urban socio-environmental flipside. In addition to asking questions about "the right to the city" (Harvey, 2008) or about "who owns the future city" (Sadowski, 2020), we need to more explicitly address the closely related question about "the right to the peri-urban". We will return to this point in the concluding chapter.

1.3.2   Understanding the Peri-Urban: A Diversity of Frames of Reference and Approaches

As discussed at the beginning of this chapter, a considerable share of future urban growth worldwide will occur in the spaces of urban expansion that can be described as "peri-urban". In the most general sense this term refers to processes of "becoming urban" (Leaf, 2011). Although there is no consensus definition (see Narain & Nischal, 2007), the term has been used mainly in three different ways: as a place, as a process or as a concept. While a detailed exposition of the connotations and usages of the term and the problems with spatial definitions of the peri-urban is beyond the scope of this chapter (for a review, see Singh & Narain, 2020), it is important to note that the peri-urban is increasingly understood in non-spatial and processual terms rather than as discrete bounded spaces. Iaquinta and Drescher (2000) were among the earlier writers to note that social and institutional contexts rather than spatial boundaries define the peri-urban. Moreover, approaches based on the assumption of clear spatial boundaries and stable states of being cannot deal with the dynamic, messy and volatile character of the peri-urban, nor with the flows of goods, resources, people and ideas across fluid boundaries. Another term often used to stress the dynamic character of the peri-urban and the flows and interactions that shape it is "peri-urban interface" (PUI) (see e.g. Simon, 2008). In this book we prefer to stress the processual characteristics of the peri-urban, but authors of the various chapters may use different terms.

In view of these developments in research and thinking about the peri-urban, approaches to the peri-urban as a bounded and recognizable spatial zone at the periphery of cities have lost their relevance. If the term "peri-urban" refers to a dynamic zone of mixed rural-urban features, encapsulated in expressions like "desakota" (McGee, 1991), then the co-existence of the rural and the urban can be found even in the heart of the city, and not just as its periphery. Moreover, the peri-urban takes on many shapes and includes processes and phenomena that cannot be expressed in exclusively spatial and geographical terms: it also exists in a socio-cultural, legal, political, institutional and economic sense. This renders a place-based definition futile. Any alternative definition will, to some extent, be arbitrary (Narain & Nischal, 2007; see OECD, 1979; Adell, 1999).

According to Friedmann (2011, p.430) "a general theory of the periurban [...] escapes us". Peri-urbanization as a process is "history in the making" (Friedmann, 2016, p.165) that "offers little scope for high-flying theorizations" (Friedmann, 2016, p.163; see also Friedmann, 2011). Despite such theoretical and definitional problems, approaching the peri-urban as a process rather than a specific type of urban region (Webster, 2011, p.632) has distinct advantages in dealing with its fluid and dynamic character. A process-based peri-urban focus emphasizes the dynamic and emergent mixes and flows of "rural" and "urban" land uses, infrastructures, economic activities, and state- and non-state institutions, identities, jurisdictions and authorities. Thus peri-urban spaces can be characterized as complex hybrid formations or socio-natural "assemblages", emergent and temporary forms of relative order and stability in highly dynamic environments reshaped by socio-natural processes and relationships (see Anderson & McFarlane, 2011; Brenner et al., 2011; McLean, 2017). This makes clear why the governing of peri-urban spaces and processes is a major problem. The dynamic character and institutional complexity of the peri-urban cannot be controlled by static structures of governance, jurisdictional boundaries, and policy institutions, while overlapping governance institutions, legal frameworks, and competing claims of legitimacy and authority are common, giving peri-urban areas their characteristic "fuzziness" (Allen, 2003; Simon, 2008; see Arabindoo, 2009) (see Fig. 1.2).

1.3.3 Peri-Urban Water Security

In disciplinary technical approaches, water security tends to be reduced to naturalized notions of water scarcity, approached through universalized and techno-managerial framings and definitions of problems and solutions, and thus depoliticized (see Joy et al., 2014; Roth et al., 2018a, b). Water (in-)security is produced in the processes of socio-natural transformation that also create peri-urban spaces, including the winners and losers that emerge in these transformations. Thus, water security is deeply social and relational, often politically contested and grounded in wider societal power structures, power relations and inequalities, (see Lankford et al., 2013; Zeitoun et al., 2016). The concept does not just refer to a technically and managerially framed "scarcity" or "water provision" but to the interplay of water access, water rights and the wider property relations around water, re-allocations

Fig 1.2 peri-urban "fuzzy" landscape, Kathmandu Valley, Nepal. (Photo Dik Roth).

and re-distributions in water use contexts that are often unequal and have a complex multi-scalar character.

Given the extremely volatile and dynamic context of the peri-urban "waterscape" (Swyngedouw, 1999; see also Budds, 2009), peri-urban water security should be researched and analysed with an awareness of its emergent and changing context, differential experiences of, and meanings given to water security, the inequalities and relations of power in such contexts, and the political, water governance and policy processes in which water securities and insecurities are produced or transformed. Following Zeitoun (2013), we argue that the main benefit of a water security focus lies not so much in using one "perfect" definition as a measuring stick for research and analysis, but rather in providing a conceptual space for interdisciplinary research of the complex interconnectedness of elements of the peri-urban assemblage. Boelens and Seemann (2014, p.1) provide a useful general description: "water security refers to people's and ecosystems' secure, sustainable access to water, including equitable distribution of advantages / disadvantages related to water use, safeguarding against water-based threats, and ways of sharing decision-making power in water governance". In view of its basically social, relational and political character, however, they prefer a plural notion of "divergent water securities" (2014, p.3).

An "integrative approach" (Zeitoun et al., 2016) to such water securities should allow for critical questions to be asked about the changing water flows and hydro-social relations, users and uses, forms of water control, access and rights, and power relations that are shaped in processes of urbanization. Whose water security gets political and policy attention? Whose knowledge and expertise, authority and influence count? Who stand to gain or loose? Who are included or excluded from water access and decision-making? To what extent are existing power relations and power differences either reproduced or transformed by the peri-urban hydro-social dynamics of the urbanization of water? How are these processes related to existing forms of social differentiation? Which discourses are used to justify and "naturalize" certain policies, courses of action, and practices of allocation and distribution?

Both urbanization and water security are closely related to the policy world, among which the Sustainable Development Goals (SDGs)[7] and, within these, especially SDG 6 (clean water and sanitation for all) and SDG 11 (sustainable cities and communities) stand out. Although we fully subscribe to these goals, one of the downsides of such bullet lists of developmental targets is that they become separated into discrete policy domains framed as unrelated problems and turned into technical solutions in ways that hide from view the basic linkages with other societal problems and development goals (such as poverty, infrastructure and climate) and, above all, with important issues like exclusion and marginalization, political participation and power, social and environmental justice, and citizenship. By zooming in on the less conspicuous dimensions of urbanization, the peri-urban cases discussed in this book clearly show some of the tensions, contradictions and dilemmas involved in processes of urban expansion and development. The relevance of a peri-urban conceptual lens to study water flows and reappropriation processes is that it challenges the neat categorizations that underpin the framing of such targets such as "sustainable cities and communities", for instance, by raising questions as what constitutes a "city" and which "communities'" sustainability we are talking about. Chap. 2 in this book (Mundoli et al.,) examines the changing access to water in peri-urban contexts in light of the move to accomplish the SDGs.

1.4 Key Themes in the Chapters: An Overview

1.4.1 The Role of Material Infrastructure

Several chapters deal with the role of water-related technology and material infrastructure, such as irrigation and drainage canals, sluice gates, and pumping devices, as key material elements in changing peri-urban waterscapes and water (in-)securities experienced by peri-urban populations. Contrary to what is often assumed, infrastructural devices like canals or water division structures are not neutral "things". Their role can be better understood from a social-constructivist perspective in which they are seen as hybrid socio-technical elements (Pinch & Bijker, 1984) designed, constructed, managed and used through often complex social-institutional processes of water control. The socially constructed character of water

---

[7] See https://www.un.org/sustainabledevelopment/sustainable-development-goals/

infrastructure becomes manifest in its design, in the recipes for its use, and in its social effects (for irrigation in India, see Mollinga, 2003; see also Roth & Vincent, 2013). As water rights and access, in- and exclusion, quantity and quality are crucially mediated by water infrastructure, it is around such infrastructure that the often competing human agencies expressed in negotiations, contestations and conflicts about rights, access and socio-technical control tend to occur; hence they have been called "signposts of struggle" (Mollinga & Bolding, 1996).

Once the role of infrastructure has been redefined as basically social, it also becomes possible to ask more basic questions, beyond its "thingy" properties. Where infrastructures exclude people physically and socially, dispossess them of their land and water, destroy their livelihoods and do other damage to their life-worlds, Rodgers and O'Neill (2012) stress that these processes can be analysed as forms of "infrastructural violence". As Ferguson (2012, p. 559; cited in Rodgers and O'Neill) argues:

> The violence that is built into the massive inequalities that dominate our societies today is often naturalized, made invisible, or made to seem inevitable, by the walls, pipes, wires, and roads that so profoundly shape our urban environments [...]. Who, then, is responsible for such violence – violence that assuredly takes lives, but in ways that seem attributable less to specific acts or agents than to […] 'a faceless set of fleeting social connections'.

Whether infrastructure exerts such violence or not, and does so directly or indirectly, its design, management and uses raise basic questions about rights (e.g. water rights; the right to sufficient quantities of clean water; protection against water), agency, causality and the allocation of responsibility for anonymous "natural" processes.[8] It also points to issues of "spatial justice" in "the social production of urbanized space" (Soja 2010, pp. 6–7; in Rodgers & O' Neill, 2012), and water justice (e.g. Boelens et al., 2018).[9]

A clear example of how peri-urban changes influence access to irrigation water and irrigation management practices is presented in the chapter by Anushiya Shrestha and co-authors. In Kathmandu Valley, urbanization is creating opportunities for new and intensified forms of exploitation, resource extraction, and opening new urban markets for water, extracted resources like sand and gravel, bricks, and agricultural produce. Through the same processes, existing (irrigated) agricultural practices and related infrastructures come under pressure. The authors discuss the gradual decay of the Mahadev Khola Rajkulo, a traditional surface irrigation canal system in Kathmandu Valley, against the background of rapid urbanization, population growth and declining water availability of this peri-urban space of Nepal's capital. Historically the stream-fed surface irrigation canals of the valley, known as *rajkulo* ("royal canals"), played a key role in local and regional food production. Configurations of land and water rights were relatively clear and stable, with water

---

[8] Another concept that can be used to analyse these processes is "slow violence". While it has been mainly used to analyse the gradual and often invisible workings of environmental pollution and the inequalities of their distribution, it can also be related to processes of urbanization and their place-based and unequally distributed forms of harm (see Nixon, 2011; Davies, 2019).

[9] For violence and development, see Escobar (2004).

rights strongly based in irrigators' contributions to canal management and maintenance ("hydraulic property"; see Coward, 1980). But this changed radically when urbanization took off, leading to inflow of population, massive land conversion and a growing pressure on water. Tracing the history of changes in canal use and management against the background of these broader transformations, the authors explain the gradual decline of this canal system, the changes in water rights, access and security associated with it, and the wider implications for canal-related cooperation and conflict. The issue discussed in this chapter also raises basic questions about ways to engage with the peri-urban: is it a rearguard fight, or can alternative (peri-)urban futures be imagined? We will return to this question in the concluding chapter.

Our second example concerns the role of wastewater canals in peri-urban agriculture. There has been a growing body of scholarship on irrigation canals in India and the water-related organizing practices, negotiations and conflicts among various (groups of) users (e.g. Mollinga, 2003). Less is known about the interactions of peri-urban water users with wastewater canals (for exceptions, see Narain & Singh, 2017; Vij et al., 2018). While there is growing attention to wastewater in urban settings (e.g. Karpouzoglou & Zimmer, 2016) and recognition of the potential value of wastewater in peri-urban agriculture —in particular in relation to its role in raising smallholder incomes and as a safe way of disposing of urban waste— little attention has been paid to the social-institutional dynamics around wastewater. The chapter by Pratik Mishra and Sumit Vij deepens our understanding of this peri-urban side of urban metabolism, especially how its effects are experienced by peri-urban communities. It describes their day-to-day interactions with and uses of these canals, the new collectivities and forms of social organization shaping up around these new sources of irrigation water, as well as the potential conflicts. Further, by juxtaposing this analysis with the effects of climate change, they contribute to a growing body of scholarship (for a review, see Narain & Prakash, 2016) on the combined effects of urbanization and climate change on peri-urban spaces, especially on their water security, noting them to experience "double exposure" (Leichenko & O'Brien, 2002).

Our last example focuses on the struggles around a sluice gate. Mohammad Shah Alam Khan and co-authors discuss conflicts around a sluice gate in peri-urban Khulna, Bangladesh, and experiences with stakeholder approaches to solve them. The location of this sluice gate —close to sea level in an area heavily affected by impacts of climate change, such as sea level rise, drainage problems and increasing salinity — makes for an extremely complex and conflict-prone water management setting in which freshwater and saline water, solid waste and wastewater, and storm runoff have to balanced. As this sluice gate has originally been designed and was operated in a way that primarily met urban interests and requirements, and not those of peri-urban inhabitants, it became another "signpost of struggle" between various powerful and less powerful urban and peri-urban interest groups competing for control over this piece of infrastructure to increase their water security. The authors describe how conflicts, political alliances and forms of cooperation arose around the gate, operation of which prioritized wastewater discharge at the expense of peri-urban fisheries, agricultural and other livelihoods. Climate changed-induced salinity intrusion and high-intensity rainfall are further complicating factors in the action research and capacity development efforts of the team that attempted to solve these conflicts through dialogue and negotiation — with uncertain outcomes.

1.4.2 Property Transformations and Declining Commons: Capturing the Urban Metabolism and Ecological Footprint

The fate of common pool resources, common property and communally managed water bodies has a special place in peri-urban studies. Processes of enclosure, accumulation, dispossession and privatization —major threats to common resources and common or communal forms of property— are often the very basis of urbanization (see e.g. Swyngedouw & Heijnen, 2003). The demise of rural and peri-urban commons in processes of urbanization in India has been analysed for expanding cities like Gurgaon (e.g. Narain & Singh, 2017; Vij & Narain, 2016; Narain & Vij, 2016). Changing uses of both groundwater and surface water, and related property and access transformations, have been studied for Kathmandu Valley in Nepal (e.g. Shrestha et al., 2018; see Fig. 1.3).

The commons are known to have several functions, including livelihood support functions. As early as the 1980s, Jodha (1986) pointed out that the commons are important for small and marginal farmers, as well as for landless households who do

Fig. 1.3 Commercial groundwater exploitation, Kathmandu Valley. (Photo Dik Roth)

not have much by way of private assets to support their livelihoods. Water commons such as lakes, ponds and wetlands also have ecological and biodiversity support functions. Many of the region's wetlands are notified as IBAs (Important Birding and Biodiversity Areas). The loss of the commons in the wake of urbanization, either through their encroachment, privatization or state takeover needs to be seriously questioned: the state, while acquiring the commons for urban expansion, ignores the multiple functions performed by them. They are often seen simply as resources for urban expansion or as dumping grounds for urban waste. These transformations of the commons are further known to deepen social inequalities (Vij and Narain, 2016). There is a need for a drastic departure from this approach, one that still needs sustained policy advocacy.

The chapter by Seema Mundoli and co-authors takes a lake and a wetland each as units of analysis for studying peri-urban water insecurity. These two cases show how the peri-urban can serve as a conceptual lens for studying urban metabolism and the manner in which the ecological footprint of urbanization is borne by the populations of peri-urban spaces. The authors show how wetlands and lakes represent two different kinds of commons that witness a compounding of stresses affecting both the quality and the quantity of the resource. This is a subject of great concern for scholars of the peri-urban, as well as for planners and policy-makers: it presents grave equity implications for peri-urban communities whose livelihoods are compromised as they lose access to the commons. In cases where communities still depend on the commons, there is a need for sustained awareness and policy advocacy to protect these communities and their livelihoods. However, it is necessary to refrain from blanket prescriptions to protect them, as over time the communities' relationships with the commons may have been or be undergoing changes as well, as Singh and Narain (2019) demonstrate. The important question, then, is: given its different meanings and ambiguities of the concept, and ongoing changes in and pressures on common resources, what "commons futures" are realistically imaginable (see also Bakker, 2010)?

1.4.3 Socially Differentiated Access to Water

Hydro-social and political ecology perspectives on water use and allocation draw attention to the relationship between flows of water and social relations of power (e.g. Swyngedouw & Heijnen, 2003). It is argued that environmental processes cannot be studied in isolation from the social and political contexts and the transformations in which they are embedded, or rather: that mutually constitute each other, empowering some and disempowering others in multi-scalar socio-ecological processes. The peri-urban, with its rich socio-economic diversity and intensifying metabolic relations with the city, is a fertile ground for studying these relationships. The existence of a high degree of social and economic diversity and heterogeneity means that there are wide variations in access to water as well. A wide variety of institutions – locally embedded norms, practices and codes of conduct — shape the differential access to water.

Recent writings on the peri-urban (see, for instance, Roth et al., 2018a, b; Shrestha, 2019) question the notion of "community resilience" in peri-urban contexts. This critique focuses, among others, on the notion of the community as a mythic, homogeneous and coherent whole. The actually existing high degree of social and economic heterogeneity, diversity and inequality in the peri-urban makes "community resilience", the latest and highly influential conceptual fashion in development policy, sound clichéd (see Kaika 2017). Analysis of water security, peri-urban or elsewhere, requires a socially differentiated analysis of access to water. There can be many axes of social differentiation: gender, age, caste, class, ethnicity, residential and occupational status – all of which intersect to co-shape access and rights to water, water-related tasks and water security.

Gender has since long been recognized to be one such axis (see Fig. 1.4). Despite this, little is known on changing gender relations around water in peri-urban contexts (for exceptions, see e.g. Narain & Singh, 2019; Vij and Narain, 2016). The chapter by Deepa Joshi and co-authors provides a situated analysis of the gendered access to water in peri-urban contexts. The authors use the case of female ready-made garment factory workers in Bangladesh to show the embeddedness of water access in wider social and power relations. They describe the daily struggles of such female workers in accessing water. Contrary to received development wisdom, their engagement with the garment factories has not resulted in "empowerment"; rather

Fig. 1.4 water and gender in peri-urban Hyderabad. (Photo Dik Roth).

it has brought increased psychological burdens of balancing home with work and living in inhospitable conditions under continuous threats of eviction and hikes in room rents.

The chapter by Nathaniel Dylan Lim and Diganta Das shows wide inequality in access to water between the core areas of the city of Hyderabad and its peripheries. Hyderabad, a growing city of South India, has appropriated water from a multitude of sources to meet the needs of its inhabitants, creating deprivation for those living at the periphery. Indeed the peri-urban is a context in which social differences translate into wide variations in access to water and water security.

1.4.4 Intervening in the Peri-Urban

The peri-urban has been described as a space "crying out for attention" (Halkatti et al., 2003). The core of the peri-urban problematique is that, as cities expand, urban planners and policy-makers focus attention on meeting the needs of the city, while neglecting the peri-urban and the resource-related needs of its inhabitants. The appropriation of land and water to meet urban needs leads to a loss of resource access and livelihood opportunities for peri-urban communities. Thus the study of the dynamics of peri-urbanization raises very basic questions about the politics of urban expansion: on what kind of assumptions are ideals, visions and policies for urban futures based? For whom are modern cities meant, planned and designed? How are the burdens and benefits of urbanization distributed, across the boundaries of the city itself? What is the role of the increasingly powerful "agents of globalization-oriented change", such as landowners, investors, real estate developers, and multinational and domestic corporate actors? (Shatkin, 2014, p.3; see also Narain & Singh, 2017). As Shatkin rightly argues, these questions basically concern issues of agency, power and social change.

There is a strong case for organized efforts to protect natural resources and the livelihoods based on them in peri-urban contexts, while at the same time building communities' capacities to demand change and to build the accountability of state institutions. Many such efforts have been made in the region in the past. These revolve mainly around community mobilization and participatory action planning approaches (Dahiya, 2003). Narain et al., (2020) describe an approach in peri-urban Gurugram in Northwest India that sought to improve local access to water by bringing peri-urban communities into direct contact with the state agencies responsible for water supply. Through a series of workshops, this led to the creation of mutual accountability relationships between the state and water users and the steering away from what Wade (1988) has described as a prisoners' dilemma situation in water.

Intervention in the peri-urban is, however, not without problems. The institutionally dynamic and heterogeneous character of the peri-urban tends to create institutional "gaps", or rather legal-institutional overlaps, pluralities and complexities between state and non-state institutions (rules, norms, institutionalized practices, codes of conduct) and practices of resource use. This has important consequences for the governance and management of peri-urban resources like land and water, often involving conflicting claims (Allen, 2003; Allen et al., 2006; Simon, 2008). As peri-urban resource exploitation related to urban metabolic processes is uneven and inequitable (Swyngedouw & Heijnen, 2003; for a recent peri-urban focus, see Bartels et al., 2020), issues of agency, power and political representation should be given due attention. If policies and interventions do not pay sufficient attention to the inequalities and forms of social differentiation emerging in peri-urban transformations and urbanization-related metabolic processes (Swyngedouw & Heijnen, 2003; for a recent peri-urban focus, see Bartels et al., 2020), they are destined to become part of the problem rather than the solution.

The growing number of peri-urban resource claimants, users and uses, and the concomitant appropriation and re-allocation of peri-urban natural resources are creating a scope for growing competition, contestation and conflict, but for new coalitions, social networks and opportunities for forms of cooperation as well (e.g. Vij et al., 2018). Such resource-related problems are often approached through the creation of stakeholder platforms and forums for policy dialogue and conflict resolution. The creation of peri-urban forums that bring diverse actors together to negotiate peri-urban issues can be the key to creating greater awareness on them, and provide a base for policy advocacy. This can only succeed if all actors involved see such issues for what they are: political. The dangers of depoliticization of basically political issues, conflicts and processes are always present in approaches based on stakeholder participation, awareness-raising and creation of community resilience (see also Kaika, 2017).

Two chapters in this volume deepen our understanding of approaches to interventions in peri-urban contexts. The chapter by Sharlene Gomes describes the use of a negotiated approach and participatory institutional analysis in peri-urban contexts in Khulna, Bangladesh and Kolkatta, India. It focuses on efforts at building the capacity of communities to talk with state agencies and demand changes. Experiences with two approaches are explored: the Approach for Participatory Institutional Analysis (APIA) for problem diagnosis and strategy exploration in problem-solving; and the Transformative Pathways, based on the Adaptation Pathways approach to planning, through which actors can explore longer-term policy strategies for sustainable peri-urban water management in a dynamic and uncertain peri-urban context.

The chapter by Mohammad Shah Alam Khan and others describes efforts at resolving conflicts between competing resource users in peri-urban Khulna through capacity development of peri-urban communities to facilitate dialogue, negotiation and conflict mitigation around an important sluice gate (see also above). This chapter demonstrates, in particular, the importance of an element of continuity and persistence in addressing these issues: the project team gained from being associated with three projects over a span of a decade, which created trust and ensured complementarity and continuity. However, the small steps forward in these processes remain time-consuming, sensitive and also uncertain in terms of their outcomes, as shown by developments in sluice gate operation.

Both chapters demonstrate that socio-economic differences, which tend to be very high in peri-urban contexts, are key in shaping the outcomes of such practically oriented efforts. These experiences also show that interventions in peri-urban contexts need to be directed at multiple scales and levels: altering power relations between the agencies of the state and water users, and altering power relations among claimants and users themselves, both urban and peri-urban. The two chapters provide fascinating accounts of these experiences and lay ground for further action research. Needless, to say, this is an area where more concerted approaches and initiatives will be needed in the years to come.

1.4.5 Conflict and Cooperation

With natural resources under growing pressure worldwide, it is not surprising that issues of resource-related conflict and questions how to turn such conflict into cooperation have topped the development agenda since the 1990s. While this theme has long been approached from a resource-deterministic perspective that assumes a direct relationship between "scarcity" and "conflict", in recent years more nuanced approaches to conflict and cooperation have emerged (see e.g. Bavinck et al., 2014). More nuance has also meant a critical reconsideration of mainstream "post-political" notions of conflict as "bad" and "cooperation" good, of a preference for "collaborative" and "participatory" approaches to those that take the politically contested character of resource conflicts as point of departure for exploring the value of conflict and dissent in processes of more radical societal transformation (Dean, 2018; Kaika, 2017; Swyngedouw, 2009). Although the contributors to this book do not directly engage with these debates, several of them engage and struggle with issues of conflict and cooperation, also making different choices in engaging with the problems in their society.

Mundoli and co-authors specifically mention issues of waste dumping in land-fills near water bodies (e.g. lakes) and grazing land, which raise questions of social and environmental justice. These and other problems involving conflicting interests are, of course, conflict-sensitive, but can at the same time lead to new forms of cooperation in solving them. Linkages between wastewater and peri-urban livelihoods, for instance, may work out in different directions and with different combinations of conflict and cooperation. Mishra and Vij note how the presence of wastewater canals can stimulate farmers to organize and engage in collective action in new ways, devise rules for water use and canal management, and thus turn canal and wastewater into a new form of hydraulic property. It is, however, important to note that such cooperative solutions also have a social, political and environmental price and may pollute the peri-urban space to make the city look "smart".

Two contributions to this book engage with conflict more explicitly. Shrestha and other authors note that changing access, rights and use practices around the irrigation canal eroded existing practices of negotiation and conflict solving. At the same time, the existence of competing interests in land and water need not necessarily cause an increase in conflicts. Open conflicts around the canal are not common, and in-migrants engaging in commercial agriculture (and thus needing water) often get access to water in ways that do not arouse and at least temporarily "manage" or dampen conflicts: by leasing-in land, building social networks and good relationships with local farmers, and investing in alternative sources and technologies (groundwater; pumping from rivers; drip irrigation) by those who can afford the investments. This is part of a general trend away from the more or less "fixed" water rights associated with the canal as "hydraulic property" towards such more individualized and pragmatic forms of access (see also Shrestha et al., 2018).

Mohammad Shah Alam and co-authors describe conflicts and attempts to manage or solve them through persistent project engagements, featuring cooperation between academics, a local NGO, government agency representatives and other local stakeholders. Key to the long-standing conflicts were the different interests in, and benefits from, the sluice gate and, hence, competing demands on its control and operation. Conflict mitigation and reconciliation involved, among others, a long-standing engagement involving a neutral actor to create trust, capacity development and the creation of a platform for dialogue. Thus, forms of dialogue and collaboration led to agreements on, among others, redesigns of the gate and changes in its management. It also brought changes such as the cultivation of less water-intensive crops and a decrease in shrimp farming. However, this is also a never-ending story: the ongoing hydro-social processes require continuous investments of time and resources, especially in a region increasingly impacted by various effects of a changing climate.

As these themes show, the peri-urban is a vibrant context for studying changing water access and the intersection of various identities that shape the differentiated access to water, and to address questions of politics and power both among peri-urban communities as well as in relation to processes of urban planning.

References

Adell, G. (1999). *Theories and models of the peri-urban interface: A changing conceptual landscape*. Literature review. Draft for discussion. Peri-urban Research Project Team Development Planning Unit, University College London.

Allen, A. (2003). Environmental planning and management of the peri-urban interface: Perspectives on an emerging field. *Environment and Urbanization, 15*(1), 135–147.

Allen, A., Dávila, J. D., & Hofmann, P. (2006). The peri-urban water poor: Citizens or consumers? *Environment and Urbanization, 18*(2), 333–351.

Anderson, B., & McFarlane, C. (2011). Assemblage and geography. *Area, 43*(2), 124–127.

Angelo, H., & Wachsmuth, D. (2014). Urbanizing urban political ecology: A critique of methodological cityism. *International Journal of Urban and Regional Research, 39*(1), 16–27.

Angelo, H., & Wachsmuth, D. (2020). Why does everyone think cities can save the planet? *Urban Studies, 57*(11), 2201–2221.

Arabindoo, P. (2009). Falling apart at the margins? Neighbourhood transformations in peri-urban Chennai. *Development and Change, 40*(5), 879–901.

Bakker, K. (2010). *Privatizing water. Governance failure and the world's urban water crisis.* Cornell University Press.

Bartels, L. E., Bruns, A., & Simon, D. (2020). Towards situated analyses of uneven peri-urbanisation: An (urban) political ecology perspective. *Antipode, 52*(5), 1237–1258.

Bavinck, M., Pellegrini, L., & Mostert, E. (Eds.). (2014). *Conflicts over natural resources in the global south.* CRC Press.

Boelens, R. A., & Seemann, M. (2014). Forced engagements: Water security and local rights formalization in Yanque, Colca Valley, Peru. *Human Organization, 73*(1), 1–12.

Boelens, R., Perreault, T., & Vos, J. (Eds.). (2018). *Water justice.* Cambridge University Press.

Brenner, N., & Schmid, C. (2012). Planetary urbanization. In M. Gandy (Ed.), *Urban constellations* (pp. 10–13). Jovis.

Brenner, N., Madden, D. J., & Wachsmuth, D. (2011). Assemblage urbanism and the challenges of critical urban theory. *City, 15*(2), 225–240.

Budds, J. (2009). Contested H_2O: Science, policy and politics in water resources management in Chile. *Geoforum, 40*(3), 418–430.

Clement, M. T. (2010). Urbanization and the natural environment: An environmental sociological review and synthesis. *Organization & Environment, 23*(3), 291–314.

Cook, C., & Bakker, K. (2012). Water security: debating an emerging paradigm. *Global Environmental Change, 22*(1), 94–102.

Cook, I. R., & Swyngedouw, E. (2012). Cities, social cohesion and the environment: Towards a future research agenda. *Urban Studies, 49*(9), 1959–1979.

Coward, E. W., Jr. (Ed.). (1980). *Irrigation and agricultural development in Asia. Perspectives from the social sciences.* Cornell University Press.

Cugurullo, F. (2016). Urban eco-modernisation and the policy context of new eco-city projects: Where Masdar City fails and why. *Urban Studies, 53*(11), 2417–2433.

Dahiya, B. (2003). Hard struggle and soft gains: Environmental management, civil society and governance in Pammal, South India. *Environment and Urbanization, 15*(1), 91–100.

Datta, A. (2015). New urban utopias of postcolonial India: 'Entrepreneurial urbanization' in Dholera Smart City, Gujarat. *Dialogues in Human Geography, 5*(1), 3–22.

Davies, T. (2019). Slow violence and toxic geographies: 'Out of sight' to whom? *Environment and Planning C: Politics and Space, 0*(0), 1–19.

Dean, R. J. (2018). Counter-governance: Citizen participation beyond collaboration. *Politics and Governance, 6*(1), 180–188.

Ellis, P., & Roberts, M. (2016). *Leveraging urbanization in South Asia: managing spatial transformation for prosperity and livability.* South Asia Development Matters. Washington, DC: World Bank.

Escobar, A. (2004). Development, violence and the new imperial order. *Development, 47*(1), 15–21.

Ferguson, J. (2012). Structures of responsibility. *Ethnography, 13*(4), 558–562.

Friedmann, J. (2011). Becoming urban: Periurban dynamics in Vietnam and China — Introduction. *Pacific Affairs, 84*(3), 425–434.

Friedmann, J. (2016). The future of periurban research. *Cities, 53*, 163–165.

Gururani, S. (2017). 'Designed to fail': Techno-politics of disavowal and disdain in an urbanising frontier. *Economic and Political Weekly, 52*(34), 38–46.

Harvey, D. (2008). The right to the city. *New Left Review, 53*, 23–40.

Halkatti, M., Purushothoman, M., & Brook, R. (2003). Participatory Action Planning in the peri-urban interface: the twin city experience, Hubli-Dharwad, India. *Environment & Urbanization, 15*(1), 149–158.

Iaquinta, D., & Drescher, A. (2000). Defining the peri-urban: Rural-urban linkages and institutional connections. *Land reform, 2*, 8–27.

Jodha, N. S. (1986). Common property resources and rural poor in dry regions of India. *Economic and Political Weekly, 21*(27), 1169–1181.

Joy, K. J., Kulkarni, S., Roth, D., & Zwarteveen, M. (2014). Re-politicizing water governance: Exploring water reallocations in terms of justice. *Local Environment, 19*(9), 954–973.

Kaika, M. (2017). 'Don't call me resilient again!': The new urban agenda as immunology … or … what happens when communities refuse to be vaccinated with 'smart cities' and indicators. *Environment and Urbanization, 29*(1), 89–102.

Kallis, G., & Bliss, S. (2019). Post-environmentalism: Origins and evolution of a strange idea. *Journal of Political Ecology, 26*(1), 466–485.

Karpouzoglou, T., & Zimmer, A. (2016). Ways of knowing the wastewaterscape: Urban political ecology and the politics of wastewater in Delhi, India. *Habitat International, 54*, 150–160.

Karpouzoglou, T., Marshall, F., & Mehta, L. (2018). Towards a peri-urban political ecology of water quality decline. *Land Use Policy, 70*, 485–493.

Keil, R. (2020). An urban political ecology for a world of cities. *Urban Studies, 57*(11), 2357–2370.

Lankford, B., Bakker, K., Zeitoun, M., & Conway, D. (Eds) (2013). Water security: Principles, perspectives, practices. Routledge/Earthscan.

Leaf, M. (2011). Periurban Asia: A commentary on "becoming urban". *Pacific Affairs, 84*(3), 525–534.

Leaf, M. (2016). The politics of periurbanization in Asia. *Cities, 53*, 130–133.

Leichenko, R. M., & O'Brien, K. L. (2002). Mitigation and adaptation strategies for global change: The case of southern Africa. *Mitigation and Adaptation Strategies for Global Change, 7*(1), 1–18.

McGee, T. G. (1991). The emergence of desakota regions in Asia: Expanding a hypothesis. In N. Ginsburg, B. Koppel, & T. G. McGee (Eds.), *The extended metropolis: Settlement transition in Asia* (pp. 3–25). University of Hawaii Press.

McLean, J. (2017). Water cultures as assemblages: Indigenous, neoliberal, colonial water cultures in northern Australia. *Journal of Rural Studies, 52*, 81–89.

Mehta, L., Allouche, J., Nicol, A., & Walnycki, A. (2014). Global environmental justice and the right to water: The case of peri-urban Cochabamba and Delhi. *Geoforum, 54*, 158–166.

Mollinga, P. P. (2003). On the waterfront. In *Water distribution, technology and agrarian change in a south Indian canal irrigation system*. Wageningen University Water Resources Series. Hyderabad.

Mollinga, P. P., & Bolding, A. (1996). Signposts of struggle: Pipe outlets as the material interface between water users and the state in a large-scale irrigation system in South India. In G. Diemer & F. P. Huibers (Eds.), *Crops, people and irrigation. Water allocation practices of farmers and engineers* (pp. 11–33). Intermediate Technology Publications.

Muldavin, J. (2015). Using cities to control the countryside: An alternative assessment of the China National Human Development Report 2013. *Development and Change, 46*(4), 993–1009.

Muzzini, E., & Aparicio, G. (2013). *Urban growth and spatial transition in Nepal. An initial assessment*. The World Bank.

Narain, V., & Nischal, S. (2007). The peri-urban interface in Shahpur Khurd and Karnera, India. *Environment and Urbanization, 19*(1), 261–273.

Narain, V., & Prakash, A. (Eds.). (2016). *Water security in peri-urban South Asia. Adapting to climate change and urbanization*. Oxford University Press.

Narain, V., & Singh, A. K. (2017). Flowing against the current: The socio-technical mediation of water (in)security in periurban Gurgaon, India. *Geoforum, 81*, 66–75.

Narain, V., & Singh, A. K. (2019). Replacement or displacement? Periurbanization and changing water access in the Kumaon Himalaya, India. *Land Use Policy, 82*, 130–137.

Narain, V., & Vij, S. (2016). Where have all the commons gone? *Geoforum, 68*, 21–24.

Narain, V., Ranjan, P., Vij, S., & Dewan, A. (2020). Taking the road less taken: Reorienting the state for periurban water security. *Action Research, 18*(4), 528–545.

Nixon, R. (2011). *Slow violence and the environmentalism of the poor*. Harvard University Press.

Organisation for Economic Co-operation and Development (OECD). (1979). *Agriculture in the planning and management of peri-urban areas. Volume 1: Synthesis*. OECD.

Pinch, T. J., & Bijker, W. (1984). The social construction of facts and artefacts: Or how the sociology of science and the sociology of technology might benefit each other. *Social Studies of Science, 14*(3), 399–441.

Rodgers, D., & O'Neill, B. (2012). Infrastructural violence: Introduction to the special issue. *Ethnography, 13*(4), 401–412.

Roth, D., & Vincent, L. (Eds.). (2013). *Controlling the water. Matching technology and institutions in irrigation management in India and Nepal.* Oxford University Press.

Roth, D., Zwarteveen, M., Joy, K. J., & Kulkarni, S. (2018a). Water governance as a question of justice: Politics, rights and representation. In R. Boelens, T. Perreault, & J. Vos (Eds.), *Water justice*. Cambridge University Press.

Roth, D., Khan, M. S. A., Jahan, I., Rahman, R., Narain, V., Singh, A. K., Priya, M., Sen, S., Shrestha, A., & Yakami, S. (2018b). Climates of urbanization: Local experiences of water security, conflict and cooperation in peri-urban South-Asia. *Climate Policy, 19*(sup1), S78–S93.

Sadowski, J. (2020). Who owns the future city? Phases of technological urbanism and shifts in sovereignty. *Urban Studies*. https://doi.org/10.1177/0042098020913427.

Satterthwaite, D. (2006). Small urban centres and large villages: The habitat for much of the world's low income population. In C. Tacoli (Ed.), *The Earthscan reader in rural-urban linkages* (pp. 15–38). Earthscan.

Satterthwaite, D. (2016). A new urban agenda? *Environment and Urbanization, 28*(1), 3–12.

Shatkin, G. (2016). The real estate turn in policy and planning: Land monetization and the political economy of peri-urbanization in Asia. *Cities, 53*, 141–149.

Shatkin, G. (2019). The planning of Asia's mega-conurbations: Contradiction and contestation in extended urbanization. *International Planning Studies, 24*(1), 68–80.

Shrestha, A. (2019). *Urbanizing flows. Growing water insecurity in peri-urban Kathmandu Valley, Nepal*. PhD thesis, Wageningen University, Wageningen, the Netherlands.

Shrestha, A., Roth, D., & Joshi, D. (2018). Flows of change: Dynamic water rights and water access in peri-urban Kathmandu. *Ecology and Society, 23*(2).

Simon, D. (2008). Urban environments: Issues on the Peri-Urban Fringe. *Annual Review of Environment and Resources, 33*(1), 167–185.

Singh, A. K., & Narain, V. (2019). Fluid institutions: Commons in transition in the periurban interface. *Society and Natural Resources, 32*(5), 606–615.

Singh, A., & Narain, V. (2020). Lost in transition: Perspectives, processes and transformations in Periurbanizing India. *Cities, 97*(102494), 1–8.

Soja, E.W. (2010). Seeking spatial justice. Minneapolis: University of Minnesota Press.

Swyngedouw, E. (1999). Modernity and hybridity: Nature, regeneracionismo, and the production of the Spanish waterscape, 1890-1930. *Annals of the Association of American Geographers, 89*(3), 443–465.

Swyngedouw, E. (2009). The antinomies of the postpolitical city: In search of a democratic politics of environmental production. *International Journal of Urban and Regional Research, 33*(3), 601–620.

Swyngedouw, E., & Heynen, N. C. (2003). Urban political ecology; justice and the politics of scale. *Antipode, 35*(5), 898–918.

Swyngedouw, E., & Kaika, M. (2014). Urban political ecology. Great promises, deadlock… and new beginnings? *Documents d'Anàlisi Geogràfica, 60*(3), 459–481.

Tacoli, C. (1998). Bridging the divide: Rural-urban interactions and livelihood strategies. International Institute for Environment and Development (IIESD), Sustainable Agriculture and Rural Livelihoods Programme, Gatekeeper Series no. 77.

United Nations (2015). *World urbanization prospects: the 2014 revision*. Department of Economic and Social Affairs, Population Division. New York: United Nations.

United Nations (2019). *World urbanization prospects: the 2018 revision*. Department of Economic and Social Affairs, Population Division. New York: .

Vij, S., & Narain, V. (2016). Land, water and power: The demise of common property resources in peri-urban Gurgaon. *Land Use Policy, 50*, 59–66.

Vij, S., Narain, V., Karpouzoglou, T., & Mishra, P. (2018). From the core to the periphery: Conflicts and cooperation over land and water in peri-urban Gurgaon, India. *Land Use Policy, 76*, 382–390.

Wade, R. (1988). The management of irrigation systems: how to evoke trust and avoid prisoners' dilemma. *World Development, 16*(4), 489–500.

Webster, D. (2011). An overdue agenda: Systematizing East Asian peri-urban research. *Pacific Affairs, 84*(4), 631–642.

Zeitoun, M., Lankford, B., Bakker, K., & Conway, D. (2013). Introduction: A battle of ideas for water security. In B. Lankford, K. Bakker, M. Zeitoun, & D. Conway (Eds.), *Water security: principles, perspectives, practices* (pp. 3–10).

Zeitoun, M., Lankford, B., Krueger, T., Forsyth, T., Carter, R., Hoekstra, A., Taylor, R., Varis, O., Cleaver, F., Boelens, R., Swatuk, L., Tickner, D., Scott, C., Mirumachi, N., & Matthews, N. (2016). Reductionist and integrative research approaches to complex water security policy challenges. *Global Environmental Change, 39*, 143–154.

Chapter 2
A New Imagination for Waste and Water in India's Peri-Urban Interface

Seema Mundoli, C. S. Dechamma, Madhureema Auddy, Abhiri Sanfui, and Harini Nagendra

2.1   Introduction

Cities have always been viewed as incubators for enterprise and innovation. However, in this urbanisation era we seem to suffer from a lack of imagination on how to handle environmental problems associated with expanding cities. The simplest strategy often is to shift the problems away from the city to its periphery or the peri-urban interface (PUI; see Chap. 1). Thus the PUI ends up having a contentious relationship with the city's core. In this chapter we look at the complexity in linkages between water and waste in the PUI of two metropolitan cities, Bengaluru and Kolkata, in India. We examine the contestations between waste and water in the PUI of these expanding cities, and the challenges that even potentially positive linkages between waste and water are faced with in cities in India.

Sustainable Development Goal (SDG) 11 on sustainable cities and communities has as one of its targets the supporting of "positive economic, social and environmental links between urban, peri-urban and rural areas [...]" (UNDP, 2015). However, the ecological footprint of cities, which measures how much productive land and water is required to sustain cities in terms of resources needed and for assimilating city wastes (Rees & Wackernagel, 1996) raises immediate concerns about attaining this SDG target—especially in the context of the peri-urban. Ecosystems, whether around or far from cities, are faced with incredible pressure when it comes to meeting a city's needs. While appropriations of resources from distant ecosystems, for example from rural areas, are considerable in any city across

S. Mundoli (✉) · C. S. Dechamma · A. Sanfui · H. Nagendra
Azim Premji University, Bengaluru, Karnataka, India
e-mail: seema.mundoli@apu.edu.in; dechamma.cs15@apu.edu.in; avisanfui@gmail.com; harini.nagendra@apu.edu.in

M. Auddy
Gramin Vikas Vigyan Samiti (GRAVIS), Jodhpur, India
e-mail: madhureema@gravis.org.in

V. Narain, D. Roth (eds.), *Water Security, Conflict and Cooperation in Peri-Urban South Asia*, https://doi.org/10.1007/978-3-030-79035-6_2

the globe (Folke et al., 1997), they are not immediately visible when it comes to their use by cities. They are, therefore, easily ignored in planning for sustainable cities. This, however, is not the case with ecosystems in the PUI, which are found in the immediate vicinity of cities. Yet, the PUI too falls through in planning practices for sustainable cities (Simon & Adam-Bradford, 2016).

The PUI referred to here is not just a physical space, but is characterised by the relationships and processes—ecological, economic, and social—that link it to the city and determine the nature of relationship with the city (Narain and Nischal, 2007). In physical terms, an important characteristic of the PUI is its mixed urban and rural land uses that support a wide variety of livelihoods: from fully agricultural to a mix of agriculture and small and medium enterprises (Mundoli et al., 2015). The PUI also provides resources to the city that it surrounds, enabling the city's development, but often at the cost of the environment of the PUI (Simon, 2008). For example, quarrying to supply construction material for the city impacts the local environment and the health of local residents (Mundoli et al., 2015). The PUI may provide affordable housing for a burgeoning workforce, which comes to the city in search of employment. Coming as migrants with different socio-economic backgrounds, many cannot afford to live in the expensive city core. The PUI accommodates the influx of population that needs basic services like water, sanitation, transport, and roads. To make matters worse, the PUI becomes a receiver of the wastes generated by the city (Mallik, 2009; Parkinson & Tayler, 2003). Thus the PUI is where the negative ecological footprint of the city first extends into; the PUIs of Indian cities are no exception (Narain et al., 2014; Shaw, 2005).

One of the main tensions in terms of access to and quality of water is between that of water and waste, both solid and sewage. According to Doron and Jeffrey (2018: 43), "never in history have so many people had so much to throw away and so little space to throw it as the people of India in the second decade of the twenty-first century". In urbanising India, with its aspirational population, this includes huge quantities of solid waste and sewage that needs to be disposed safely. Unfortunately, our cities have not been able to achieve even proper segregation of waste at source, or in recycling or reusing. Thus the bulk of the urban waste tends to be dumped in landfills (Ministry of Urban Development, 2016b). These landfills are often located in the PUI of cities, posing a threat to the environment and the people who live there. Waste is also dumped into water bodies in and around cities. Sewage from both households and industrial effluents ends up in water bodies too; some of it has been treated but the bulk remains untreated. These water bodies, receivers of solid waste and sewage, are at the same time accessed for household and livelihood needs by urban, peri-urban and rural communities (ESG, 2018; Narain et al., 2014; Parkinson & Tayler, 2003). The interlinkages between waste and water in Indian cities, then, are extremely complex and can have a range of impacts.

In this chapter, we look at water and waste in two sites in the peri-urban spaces of two state capitals in India: Lake Kannuru in the periphery of Bengaluru city, situated in the south-Indian state of Karnataka; and villages in *mouzas* falling in the peri-urban wetlands east of Kolkata situated in the state of West Bengal in eastern

India.[1] Water systems in both areas support the livelihoods of local communities and migrants. But while the freshwater of Lake Kannuru supported local livelihoods and subsistence, in the case of the wetlands in Kolkata sewage from the city was diverted for agriculture and aquaculture. We unravel the complexity in the waste-water relationship through these comparative cases, in which city waste is seen as a pollutant in one case and as a nutrient in the other. The chapter is an exploration into the complex linkages between waste and water that the PUI of cities face and that are a challenge to attempts to move towards a more sustainable urban future for India. We also present the potential and contestations in reimagining these linkages between waste and water in expanding cities from an open to a closed loop system (Smit & Nasr, 1992).

The chapter begins with a brief outline of the interlinked social and ecological landscape of the two cities, and includes a description of the specific study sites. In the next section we describe the methods, which is followed by the findings from the research. The findings include some of our inferences. In the last section we discuss the continuing challenges with regard to the waste and water interlinkages in the urban future projected for India.

2.2   Study Area

Bengaluru (earlier known as Bangalore) has its origins as a medieval town in the sixteenth century, though there are records of settlements in the region dating as far back as 6000 years, to the Stone Age. The city is situated on the Deccan Plateau in the southern Indian peninsula in a semi-arid region. The disadvantage of not having a perennial river to supply water was overcome by settlements in this landscape by constructing lakes to capture rainwater. The undulating topography enabled construction and diversion of rainwater into depressions, with a network of channels connecting upstream lakes to the ones downstream. Early epigraphic inscriptions also provide proof of livelihoods around agriculture and livestock rearing (Nagendra, 2016). The construction of lakes continued even after the initial boundaries of the town were drawn in 1537 CE. Towards the end of the eighteenth century, the city came under colonial rule. Since then lakes have had a mixed fortune. While the British saw the revival of lakes as important for boosting agriculture to increase their revenue, the demographic and spatial expansion of the town, as well as changing perceptions about the lakes, have had their impacts. Lakes and their wetlands began to be either seen as sites of disease, or prioritised for aesthetic purposes (Unnikrishnan et al., 2016). They were also increasingly converted to other forms of land use. The city gradually looked to more distant sources of water to meet its requirements, thus further reducing their importance. In the post-independence

---

[1] *Mouza* refers to an administrative unit in West Bengal containing one or more settlements or villages.

period there was a decline in the number of lakes in Bengaluru (Thippaiah, 2009). More recently, with rapid urbanisation, lakes have also become filled with sewage and industrial effluent (Nagendra, 2016). Many lakes in the city core, and increasingly also in the PUI into which they city is expanding, have been adversely impacted (Mundoli et al., 2017). The PUI has also become a dumping site for urban waste, and this too has had impacts on the environment. Once known as the "City of Lakes", Bengaluru is now called the "Garbage City" (Nagendra, 2016). The city generates huge amounts of garbage and sewage, but has failed to scientifically manage it. Waste is dumped in peri-urban villages, compromising the environment, including that of the lakes, and the health of local residents.

At present the unsegregated waste from the city is dumped in an abandoned quarry bordering Kannuru village, one of our peri-urban study sites situated northeast of Bengaluru city. It is one of the eight villages in Kannuru *Gram Panchayat*.[2] The area of Lake Kannuru, situated northwest of the village, is 25.92 ha. Kannuru has been a site of longitudinal research for us since 2013 (Mundoli et al., 2015). In this chapter we look at the changes to the lake and the environment of the village, and especially the impacts on livelihoods around the lake and the lives of local residents since the area adjacent to the lake was converted to a landfill.

Our second study site, the wetlands situated in the PUI east of Kolkata (earlier known as Calcutta), contrasts with the first one. Kolkata had its origins in a small cluster of villages situated on the eastern banks of the Hooghly River, a tributary of the Ganges. Detailing the events that led to the formation of Kolkata City since the setting up of a trading post of the East India Company in 1690 is beyond the scope of this paper. Suffice to say that, unlike Bengaluru, which is situated in a semi-arid region and grew around its human-made lakes, Kolkata originated on the banks of a perennial river, which influenced its growth. The city grew eastwards from the banks of the Hooghly by reclaiming marshy areas, which were seen as inhospitable but necessary to be reclaimed for expansion (Ghosh & Sen, 1987).

While the location of Kolkata was viewed as beneficial for trade by the British, the climate and the poor sanitation of the growing town were a cause of illness and mortality among the European and Indian population (Deb, 1905; Martin, 1839). As the population of the city grew, water-borne diseases became a particular concern. Although situated on the banks of a river and dotted with numerous tanks and wells, potable water continued to pose a problem for both the British and the Indian population. In the 1840s a project for supplying purified water through a piped system across the city from the Hooghly River was mooted. Though expensive, the British went ahead with the project, thereby adding "a new milestone in the journey of Calcutta towards urbanity (De, 2014: 291)".

But then there was the issue of sewage, including human waste, and garbage generated by the growing city. In his famous 1803 min, Lord Wellesley had indicated that it might have been an error to drain the city westwards into the Hooghly, while the incline of the city, though very slight, was more towards the southeast.

---

[2] A *Gram Panchayat* is the formal local self-governance system at the village level in India.

One of his recommendations was to ascertain in which direction the drains and water courses needed to be altered (Martin, 1836). In the 1850s, William Clark came up with a scheme for Kolkata's drainage with the salt lakes to the east identified as an outfall. The scheme was sanctioned and completed in 1884 (Bunting et al., 2005; Mitra, 1990). Meanwhile in 1865, one of the sites identified for disposal of the waste of the city was Dhapa, east of the city. The land, covering a square mile (the "Dhapa Square Mile") was leased to an individual who began to undertake farming on the garbage substrate in the 1880s. The tidal River Bidyadhari that lay further east of Kolkata in the 1850s, served as a source of brackish water for fish cultivation. After this river had been declared dead in 1928, sewage from the city was considered to serve as a substrate for fish cultivation (Ghosh, 1988; Ghosh & Sen, 1987). The sewage continues to flow into these wetlands, and sewage-based agriculture and aquaculture are recognised as an example of how peoples' knowledge has contributed to the productive use of urban waste for livelihood (Ghosh & Sen, 1987).

However, over the years the quantity and quality of sewage has altered, as have land use and land cover around the wetlands. Reclamation of the salt lakes has been proposed since the 1830s, but it was only in the 1960s that the reclamation of large areas actually began (Ghosh & Sen, 1987). In the 1990s, out of concern with the increasing reclamation of the wetlands and conversion to built space, a public interest litigation was filed seeking protection for the wetlands. The resulting court order demarcated the wetland for protection. This was strengthened in 2002, when 12,500 ha of the wetlands, was demarcated as the East Kolkata Wetlands (EKW), and also gained international recognition as a protected Ramsar site. The wetlands also received attention thanks to the efforts of Dhrubajyoti Ghosh, a sanitation engineer with the West Bengal government who fought to protect the wetlands and the livelihoods of people who depended on it (Calcutta High Court, 1993). The wetlands acquired a unique status because, unlike other Ramsar sites, sewage-based agriculture and aquaculture was practised on a large scale here, contributing to local livelihoods. Villages in the wetlands that fell within the boundary of Ramsar-demarcated East Kolkata Wetland were our second study site.

The *mouzas*, where the study villages are situated, lie in a landscape dominated by water. *Bheris* (fishponds) of different sizes—from small ones owned by individual fishermen to others covering hundreds of acres and run by cooperatives—spread across the wetland. Agriculture is practised as well, along with raising livestock and kitchen gardens where vegetables and spices are grown for household consumption. While both agriculture and aquaculture continue to be practised, using sewage water from the city, these livelihoods are witnessing several challenges. These will be discussed below.

2.3   Methods

In both sites we have conducted in-depth qualitative research, mainly through interviews, in the peri-urban landscape where water and waste issues intersect. In Bengaluru, we conducted 11 individual interviews around Lake Kannuru and the village in October 2019. Interviewees included local residents who engaged in farming and livestock rearing for their livelihoods, and the local resident who held the tender for fishing in the lake. In addition, we spoke to officials from the local government departments. Our focus here was to understand how the lives and livelihoods of local residents, especially the livelihoods linked to the lake, have been impacted by the landfill that had been set up near the village.

In the case of the Kolkata wetlands, we conducted interviews in seven villages in the *mouzas* of Bhagabanpur, Kharki, Deara, Hadia, and Tardah Kapashati located in the South 24 Parganas District adjacent to Kolkata District. For this research we conducted 23 interviews and interacted with 26 persons from the EKW in November and December 2018. We conducted a total of 22 individual interviews with farmers (8), fishers (8), and migrants (6). The interview durations extended between 35 and 50 min. Of the individual interviews, seven were done in Bhagabanpur, nine in Kharki, and two each in Deara, Hadia and Tardah Kapashati. The group discussion was conducted in Deara with four individuals—two fishers and two farmers—and this extended for a little over an hour.

Our focus here was to understand the challenges faced by people engaging in fishing and farming, both of which are livelihoods that are closely linked to the quality and quantity of sewage that the wetlands receive from Kolkata city. The interviews we conducted with the migrants who had settled in the wetlands more recently, were held to see what their use of and perspectives on the wetland were and to compare the migrants' views with those of long-term residents. The time span of their residence in the wetlands ranged from 7 months to 6 years. Two of the six migrants were construction labourers working in Kolkata city, two sold jaggery for a living, one had a small shop, and another was a retiree who had recently moved to the area. We have included quotations from the responses given by the interviewees as illustrative examples and to capture the perspectives of interviewees in their own words, as they are important participants in our research (Corden & Sainsbury, 2006).

2.4   Squandering Our City's Water – And Waste

2.4.1   *Water Commons in Crisis: The Impact of Bengaluru's Landfill on Lake Kannuru*

In our research initiated in 2013, the objective had been to understand the use, management and transformation of lake, *gunda thopes* (wooded groves), cemeteries and grazing lands around Kannuru (Mundoli et al., 2015, 2017). While tracing the transformation of Lake Kannuru we found that urbanisation with changes in land use had reduced the inflow of water into the lake. One of the main reasons for this was the destruction of the hillock to the west of the lake due to quarrying. The water quality, however, had deteriorated only slightly compared to other lakes in peri-urban Bengaluru into which sewage and garbage had been dumped. The reduction in water quantity—due to the effects of stone quarrying, the increase in built-up area around the lake blocking inflow channels, and reported reduction in rainfall patterns in recent years (Mundoli et al., 2015)—had meant that traditional livelihoods had been impacted. Thus, paddy cultivation had been discontinued, though some millet and fodder grass cultivation still persisted in the wetlands east of the lake. The livestock herders continued to access the lake for washing and watering cattle, and for grazing and fodder grass collection, though the reduction in water quantity had impacted access to both. The tender holders for fishing found it hard to break even, as they had to deal with reduced catch and increased costs of pumping water into temporary shallow pools for raising fish. However, these traditional livelihoods linked to the lake were still found to persist (Mundoli et al., 2015).

By September 2016 the quarrying had ceased and mixed waste from the city began to be dumped in the abandoned quarry pits. These pits were not part of Kannuru *Gram Panchayat*, but of Bellehalli Ward, which falls within the *Bruhat Bengaluru Mahanagara Palike* boundary.[3] However, since the quarry, now converted into the landfill, lay on the boundary between Bellehalli and Kannuru, the adverse impacts of the landfill were soon felt by the residents of Kannuru. According to one of the interviewees, reminiscing about the past use of the lake:

> You know … the lake looked white in colour from far, that's how pure it was. When we would come to graze…when the cattle would drink water, we also put our head down and drank water like them … along with them.

This lake is in a very different state today. The water has turned black in colour and gives off a pungent smell. According to local population, the blasting for quarrying had resulted in cracks being formed in the quarry walls. As a result, leachate from the quarry pits filled with waste escaped into the lake. The farmers and herders provided perspectives on how the landfill had affected their livelihoods. One of the interviewees, a woman grazing her cattle adjacent to the lake, had migrated to Kannuru from a village in the neighbouring state of Tamil Nadu. Escaping drought, she and her husband had settled here 10 years ago to work in the quarry. Even after quarrying was stopped she, her husband and their three children continued to live in the hutment on the northern boundary of the lake, close to the landfill. She had sheep and goats of her own and, for a fee, also herded cattle of villagers around the lake. According to her, the water quality of the lake was very bad, especially during rains, when the water from the quarry overflowed into the lake. While she accesses water from a filtered source at a distance, the livestock continued to drink the filthy

---

[3] The *Bruhat Bengaluru Mahanagara Palike* is the administrative body responsible for providing civic amenities and infrastructure to the Greater Bengaluru Metropolitan area.

water. She was, however, concerned that if this continued the livestock too would fall sick.

The livestock numbers brought to the lake for watering and grazing had also decreased considerably. At present some herders only collected fodder for stall feeding. The other danger with grazing was the increase in the number of dogs, attracted by the garbage, that chased and bit the goats and sheep. Her own household use of the lake for washing clothes and bathing had stopped when the family had begun to fall ill and develop rashes. The other herders we interviewed shared similar concerns about the lake. The main income source of one of them is the milk from his cows. Even though he lived adjacent to the lake he said that he did not collect fodder or use lake water, as he is concerned about the impact on the health of his livestock. Instead, he collects fodder from fields located at a distance. According to him, the cattle in the village have also been producing less milk than before, though he said he could not attribute this to the landfill.

The tender holder for fishing in the lake said that the water quality had deteriorated owing to pollution from dumping, resulting in around 5000 fingerlings dying. When the water quality was better, more people came to buy fish, but nowadays he is unable to make any profit from fishing. The local population had stopped buying the fish because of the pollution, claiming that the fish did not taste as good as in the past. At the time of our study, no fishing was happening in the lake under the contract, but a few local people caught fish in the lake on Sundays.

According to the interviewees, cultivation around the lake, which had started declining in 2013, had completely stopped now. Even if there was water in the lake, the channels that carried the water to the fields downstream had fallen into disrepair and were choked with weeds. Similarly, the rituals and traditions around the lake had been forgotten. Locals who used to frequent the lake avoid it now because of the stench and mosquitos. A whole way of life has changed for the village. One of the elderly interviewees who used to spend time with other residents under the shade of the *neem* (*Azadirachta indica*) and *peepul* (*Ficus religiosa*) trees in the village square said:

> "We are supposed to inhale good air under the neem tree but we can only smell garbage. When we open our doors in the morning we smell garbage. That's how our day starts."

The officials we interviewed shared their concerns about the impacts on the village too, though they were wary of revealing too much. Confirming what one of the interviewees had said regarding milk yield, one official said that the quality of milk had definitely come down. He was also concerned about the health of local residents who retained a portion of the untreated milk for household consumption. With regard to direct impacts of the landfill on human health, there were concerns of more people falling sick as a result of groundwater contamination and working in the landfill. The migrant population that lives in Kannuru is especially vulnerable, as they work in the landfill for higher wages but fall sick from viral fever, respiratory problems and skin diseases. Questions that village level officials raised regarding the impacts of the landfill were brushed aside by higher-level district officials saying that dumping was not happening within the Kannuru village boundary. But, as one of the interviewees asked:

"How can they draw boundaries to water and air? Will they be in the same place?"

This quote brings out very clearly the challenges of governance, where jurisdictions between the urban ward and rural panchayat, neatly demarcated on paper, do not account for realities in the PUI.

2.4.2   Sewage as a Livelihood Source: Impacts on Wetlands East of Kolkata

The wetlands east of Kolkata not only serve as the kidneys of the city, but also support agriculture and aquaculture. The interviewees we spoke to had been doing aquaculture and agriculture in the wetlands for a decade and beyond. They had learnt the skill from their fathers and grandfathers, who worked in the *bheris* and farms. Except for one fisher who was doing the work for 10 years, fishers and farmers had been engaged in their livelihoods for 15–45 years. The interviewees all owned their own *bheris* or farms. While this meant they could enjoy the income from these activities, working on the *bheris* and farms was hard work. In the case of the *bheris*, strengthening the embankments by growing hyacinths, dredging the *bheris* to a correct depth, mixing the correct amount of sewage as food source for the fish, introducing spawn, and mixing lime and oil cake were done by the owners themselves, sometimes assisted by sons and grandsons. Catching and selling fish in the local market was another part of their work. Only one interviewee employed labour to work in his *bheri*. In the farms, too, the work for growing vegetables was done by the farmers, who took care to see that the right amount of sewage was mixed with the soil. The fishers and farmers assured us that the quality of fish and produce from the farms were absolutely safe and they were not only sold in the markets but also consumed by themselves.

But much in the landscape of the wetlands has changed and is causing concern to the farmers and fishers alike. One of the main concerns was with regard to land use changes. Several interviewees mentioned that there has been an increase in the number of people coming to live in the wetlands. To accommodate the needs of the resident and migrant population, farms and *bheris* have been converted into houses and other infrastructure. The network of roads has also expanded, which was regarded as beneficial by the interviewees for two reasons: the roads provided connectivity to the population, many of whom worked in Kolkata; and for the fishers the roads served as barriers preventing overflow from one *bheri* to another. A specific land use change, in this case the construction of roads, that we would expect to adversely impact the livelihoods was seen as beneficial by one group—the fishers. The conversions of land were the result of sales of land by fishers and farmers. Interviewees also mentioned feeling pressurised into selling their lands by the real estate lobbies.

In addition to such impacts on land use, there have also been impacts on the quantity and quality of sewage for farming and fishing. In terms of quantity, many interviewees alleged that the canals bearing sewage were filled with debris and mud that caused siltation and reduced the flow of sewage. According to the interviewees, this was a strategy to force them to give up fishing and farming, and ultimately sell their lands. The disruption in flows of sewage meant that there were less nutrients available for the farmers. This has influenced their produce, even though they may still have access to fresh water. For the fishers, who adjusted the sewage quantity to generate fish feed, less access to sewage meant that they had to buy fish feed from the market, which increased their costs and reduced their profits.

The quality of sewage was also a cause for concern. One of the main problems, according to the interviewees, was that effluents from the leather factory situated in the wetland, albeit outside the EKW boundary, were dumped into the domestic sewage. This impacted the quality of the produce, be it fish or vegetables. The fishers we interviewed also reported fish deaths in the last 4 years, especially between October and January; dead fish were in evidence during our field visits as well. They were not exactly sure what the cause of the disease was—speculating about either changes in water quality or climate change as causes. To avoid a complete loss of income, the *bheri* owners had to buy medicines from the market that they sprayed on the water. These medicines were expensive and further reduced their profit.

According to the interviewees, making ends meet in farming and fishing had always been challenging, but with the land use changes and issues of sewage supply their problems exacerbated. For all but two interviewees, who sometimes worked as wage labourers, fishing or farming was their only source of income and only skill. As one of the interviewees said:

> We, the fishermen, here depend on the *bheris* for our income to meet the daily needs of our household. From waking up in the early morning at 4 am and going to the *bheris* for fishing and then selling it in the local markets to working hard daily to maintain and protect this wetland —for us the *bheris* you see are all that we have … without the *bheris* we don't have any secure future.

Interviewees also mentioned that farming was on the decline and more farms were being converted to *bheris*. This is an interesting insight that the interviews have yielded to date, but more research needs to be done to understand the causes of these changes. But for those who continued with farming, land was of immense importance. According to a farmer:

> "Losing the land would mean losing everything. I would not be able to feed my family then."

While some interviewees expressed that the unprofitability of their livelihoods may force them to sell their lands, others were adamant that they would fight to protect the wetlands that put food on the table for their families. According to one fisherman:

> "The *bheris* are close to our hearts … we will fight hard to protect the existing *bheris* and farms … without these we don't know where to go."

To what extent the marginalised can hold out in the face of the real estate lobby, however, is not clear. The motivations to protect the wetland are owing to direct livelihood and subsistence benefits, but this too is changing, as was seen in the case of the next generation, which is moving away from farming and fishing as source of livelihood. One of the older interviewees felt that the youth were losing the connect with the wetland, and did not have the same attachment to traditional livelihoods as their parents. The youth prefer to work as construction labourers in Kolkata or at the leather factory, where the income is much higher. This is ironical, since the fishers and farmers in the interviews said that the same factory was polluting the sewage with chemicals. But the higher wages at the factory make it an attractive option. The preference among the youth for moving to other jobs was also mentioned as a cause of labour shortage for work in the farms and *bheris*. Whether to continue with traditional activities or to migrate is a dilemma faced across peri-urban settings in India—and choices made are also very site specific (Purushothaman & Patil, 2019). The question we need to explore in a productive landscape like the EKW with the potential to support livelihoods is: are the reasons for discontinuing livelihoods purely the result of unviable farming or fishing? Or is it the result of a range of factors discussed in the paper, such as pressures on and conversion of wetlands, declining access to sewage and freshwater.

Our interviews with migrants threw up some interesting perspectives. Unlike the fishers and the farmers, the migrants we interviewed were not dependent on the wetland for their livelihoods, which may have influenced their views on the wetland. They were recent residents, and their unfamiliarity with the landscape and limited dependence on the wetland for livelihood were evident in the interviews. One of them said:

> "Before coming here I never heard of the wetlands. After coming here I was surprised to see such large areas covered with water."

With regard to the changes of the landscape, the interviewees observed that they had been here for too short a time to comment on the land use change. But, with the exception of two, all had similar views on these changes. The interviewees felt that the development-related land use changes, such as roads, buildings, houses and schools, were good, as they improve the lives of those living in the wetland. However, the wetland was seen as a garbage dumping ground, and the *bheris* and farms as a health hazard. As one of the interviewees said:

> "These are nothing but mosquito breeding grounds."

Health has historically been given as a reason for filling up and reclaiming marshes and wetlands around Kolkata. The increase in migrant population with no links to livelihoods could contribute to further conversion of the wetlands, to the detriment of the farmers and fishermen and the city itself. However, a couple of interviewees, also migrants, emphasising that development was necessary, also observed that it would negatively affect the livelihoods of farmers and fishers. The retiree we interviewed was happy that, unlike the city where he had lived, the wetlands were not polluted and he could breathe fresh air. For some interviewees the use of the wetlands was limited to washing clothes and bathing. Most of them were not concerned about the quality of fish and vegetables from the wetlands that they consumed. One of them, though, was even unaware that the fish and vegetables he consumed were grown in the wetlands.

2.5 Discussion and Conclusion

Urbanisation presents many dilemmas for attaining the sustainable development goal on sustainable cities and communities (UNDP, 2015). The rising urban population creates more demand for resources like food and water, but also increases waste generation, both sewage (human and industrial) and solid waste (wet, dry and hazardous). Urbanisation thus compromises the quantity (Mundoli et al., 2015) and quality of water sources, with untreated sewage and unsegregated garbage dumped in and around waterbodies of cities and peri-urban spaces, as shown in Kannuru. At the same time, wastewater can support livelihoods and feed the city, as agriculture and aquaculture using wastewater in the wetlands in Kolkata's PUI shows. Clearly, there is much complexity in the water-waste interlinkages in the PUI of Indian cities, and the challenge lies in ensuring that solid waste and sewage are used productively with minimal impact on ecological resources such as water.

Landfills are often found outside city boundaries (except in the case of large metros such as Mumbai and New Delhi, where city expansion has extended to encircle older landfills). In Bengaluru, landfills have moved from one site in the PUI to another. Dumping of unsegregated and untreated garbage in villages in the periphery of Bengaluru, adversely impacting the health of residents and the environment, have led to protests by local communities. The response to this by the administration has been to simply shift the landfill to another site (Environmental Support Group, 2018). The result is that today the landfill is situated adjacent to Kannuru, once again posing a hazard for the village and its residents. The rules with regard to solid waste management specify a hierarchy determining that only residual waste is to be disposed in landfills. The rest of the focus is on reduction of waste generation, reuse, recycling and converting waste to energy. The very existence of landfills like the one next to Kannuru, with tonnes of unsegregated dumped waste, is in contravention of these rules (Ministry of Urban Development, 2016a). The responsibility of the citizens with regard to protecting the environment has been laid out in the Constitution of India and has been referred to in court rulings (High Court of Karnataka, 2012). But cities in India are nowhere close to managing their waste as laid out in the legislations and rules (Ministry of Urban Development, 2016b).

Using wastewater for agriculture and aquaculture is a common practise in cities across the world (Costa-Pierce et al., 2005; Scott et al., 2004), and there are many examples in India as well. Van Rooijen et al. (2005) have explored the links between urban water supply and irrigation in the city of Hyderabad. The concerns the authors express about the impact of wastewater irrigation on health, environment and crops are mentioned as requiring additional research, but finding ways to reuse wastewater is seen as critical for expanding cities (van Rooijen et al., 2005). Yadav et al. (2016) have explored the impact of wastewater used in agriculture in peri-urban Udaipur, where the accumulation of harmful metals in vegetables irrigated with wastewater is a concern. In Hubli-Dharwad safer and sustainable farming solutions to deal with impacts of wastewater irrigation are seen as possible by creating awareness among farmers (Bradford et al., 2003).

What is common to all these studies is that, as water supply to cities increases, wastewater irrigation is seen as inevitably replacing blue water irrigation. Similarly, aquaculture has been an important source of livelihood in urban and peri-urban cities across the world (Costa-Pierce et al., 2005). In India, various technical aspects of aquaculture in the wetlands around Kolkata have been documented, especially its role as the sewage treatment plant of Kolkata (Ghosh, 1988). The aquaculture in Kolkata's wetlands, however, is also an example of how local fishermen have used their traditional knowledge to earn a livelihood; this too has received some attention (Bunting et al., 2005; Ghosh, 2005). Aquaculture in Karnal, in the state of Haryana, was different: there a large facility for aquaculture was run by the government and later leased to a private individual. This effort is seen as an effective way of optimal resource recovery, also supplying food to the urban population (Kumar et al., 2015).

Waste in landfills that pollute water bodies in the PUI and sewage that supports livelihoods are very different framings of the impacts of the waste that our cities produce. The question, then, is how to make optimal use of the waste that cities generate. As a starting point, urban wastes, both garbage and sewage, need to be seen not just as polluters of water bodies but as a "resource for sustainable development" (Smit & Nasr, 1992: 143). This requires moving from the waste in-waste out open loop to the closed loop system, in which wastes and water can be used sustainably. This will require that we look more closely at our waste and water systems and see how we can best close the loop in each of our cities (Smit & Nasr, 1992).

In spite of the opportunities and potential of using organic solid waste as a substrate for farming, and wastewater for agriculture and aquaculture in India and across the world, the positive elements of waste receive little or no attention from policy makers or government officials. Waste generated by cities is hardly seen as a resource for meeting the increasing demand for food by a growing population in the very cities that generate it (Buechler et al., 2005; Saldias et al., 2017; Scott et al., 2004). Therefore, to begin with, there needs to be a change in perception with regard to waste and water. Solid waste needs to be reduced, recycled and reused in keeping with the regulations laid down for this.

More research in the context of waste and water specific to the PUI of cities of India is required to fill the knowledge gaps. These include how organic garbage can be used effectively as a substrate for growing food supplied to the city, while ensuring that the health of producers and consumers is not compromised. We need a better understanding of perceptions and choices of farmers with regard to using wastewater for irrigation or moving from blue water to wastewater irrigation. In the case of sewage-based agriculture and aquaculture, we need to understand how fishers and farmers who have traditionally been engaged in this livelihood, as in the case of the Kolkata wetlands, have been able to adapt to changing composition and quantity of sewage. We also need regulatory frameworks to address issues of health and impact on environment (Buechler et al., 2005; Saldías et al., 2017).

It should also be recognised that PUIs are not just sites for disposal of a city's waste—and should not be seen and treated as such—but are themselves sites of production, supporting livelihoods and subsistence use. Instead of taking wastewater to far-off areas for irrigation, it is important to explore the possibilities of wastewater use closer to its source in the PUI and investigate what forms of decentralised management of wastewater would be optimal, as is being done in other cities in the Global South (Parkinson & Tayler, 2003). We need to explore innovative ways of nutrient recycling supported by policies that result in positive and sustainable outcomes for waste-water linkages in the PUI.

We need to challenge both the narratives about and the reality of PUIs in Indian cities. This will require finding ways to minimise the harmful impacts of the ecological footprint of cities, and also recognising that the city itself stands to gain by doing so (Rees & Wackernagel, 1996). Attaining the SDGs in the context of PUIs is definitely challenging for a variety of reasons. Sustainability in the PUI for us means ensuring environmental protection along with social justice. But instead of steps to bring these two goals together, there is a trend towards their divergence. This happens in different ways. For one, displacing the marginalised is often seen as necessary to protect the environment in the PUI (Marshall et al., 2009). This is visible in the case of lakes in Bengaluru, for example, where lake renovation by landscaping is initiated by wealthier urban residents along with the government. This results in alienating the urban poor, such as herders, who are dependent on these lakes (Nagendra, 2016). Dumping of the city waste sites in landfills, close to resources such as lakes and grazing lands that the peri-urban poor depend on, is clearly a concern when it comes to social justice. Second, in the changing PUI we can see instances of cooperation, conflict and conflict of interest co-existing, as in the case of peri-urban Gurgaon (Vij et al., 2018). A similar situation could be foreseen in the case of the wetlands of Kolkata, where linkages that exist between wastewater and livelihoods can shift positively or negatively, depending upon land use changes; only more research can reveal how such transformations will work out. For us, to move towards realizing the SDGs in the context of the PUI of cities it is imperative that we learn from situations of conflict and cooperation, building empirical information through research and working collectively to better protect the environment of the PUI and ensuring social justice for communities.

Acknowledgements We thank Azim Premji University for supporting the research. We acknowledge Manjunath B. for his assistance with fieldwork. We thank the interviewees who gave their valuable time and patiently answered all our questions.

References

Bradford, A., Brook, B., & Hunshal, C. S. (2003). Wastewater irrigation in Hubli–Dharwad, India: Implications for health and livelihoods. *Environment and Urbanization, 15*(2), 157–170.

Buechler, S., Mekala, G. D., & Keraita, B. (2005). Wastewater use for urban and peri-urban agriculture. In R. van Veenhuizen (Ed.), *Cities farming for the future, urban agriculture for green and productive cities* (pp. 244–260). RUAF Foundation, IDRC and IIRR.

Bunting, S. W., Kundu, N., & Mukherjee, M. (2005). Peri-urban aquaculture and poor livelihoods in Kolkata, India. In P. B. Costa, A. Desbonnet, P. Edwards, & D. Baker (Eds.), *Urban Aquaculture* (pp. 61–76). CABI Publishing.

Calcutta High Court. (1993). *People united for a better living in Calcutta vs State of West Bengal and Others*. Constitutional Write Jurisdiction, Decision 24 September 1992, Calcutta High Court, Calcutta, India.

Corden, A., & Sainsbury, R. (2006). *Using verbatim quotations in reporting qualitative social research: Researcher's views*. Social Policy Research Unit, The University of York.

Costa-Pierce, B., Desbonnet, A., Edwards, P., & Baker, D. (Eds.). (2005). *Urban aquaculture*. CABI Publishing.

De, B. (2014). The city and its scientific water project. In B. Chakraborty & S. De (Eds.), *Calcutta in the 19th century: An archival exploration* (pp. 284–291). Niyogi Books.

Deb, R. B. K. (1905). *The early history and growth of Calcutta*. Romesh Chandra Ghose.

Doron, A., & Jeffrey, R. (2018). *Waste of a nation: Garbage and growth in India*. Harvard University Press.

Environmental Support Group (ESG). (2018). *Bangalore's toxic legacy intensifies: Status of landfills, waste processing sites, dumping grounds, and working conditions of pourakarmikas*. Environment Support Group.

Folke, C., Jansson, A., Larsson, J., & Costanza, R. (1997). Ecosystem appropriation by cities. *Ambio, 26*(3), 167–172.

Ghosh, D. (1988). Wastewater fed aquaculture in the wetlands of Calcutta: An overview. *Proceedings of the international seminar on wastewater reclamation and reuse for aquaculture*, 6–9 December 1988, Calcutta, India.

Ghosh, D. (2005). *Ecology and traditional wetlands practice: Lessons from wastewater utilisation in the East Calcutta Wetlands*. Worldview.

Ghosh, D., & Sen, S. (1987). Ecological history of Calcutta's wetland conversion. *Environmental Conservation, 14*(3), 219–226.

High Court of Karnataka. (2012). *Daily orders of the case number: Writ Petition 24739/2012 for the date of order 10/09/2012*. High court of Karnataka.

Kumar, D., Chaturvedi, M. K. M., Sharma, S. K., & Asolekar, S. R. (2015). Sewage-fed aquaculture: A sustainable approach for wastewater treatment and reuse. *Environmental Monitoring and Assessment, 187*, 656, 11 pages.

Mallik, C. (2009). Urbanisation and the peripheries of large cities in India: The dynamics of land use and rural work. *Indian Journal of Agricultural Economics, 64*(3), 421–430.

Marshall, F., Waldman, L., MacGregor, H., Mehta, L., & Randhawa, P. (2009). *On the edge of sustainability: Perspectives on peri-urban dynamics*. STEPS Working Paper 35, Brighton STEPS Centre.

Martin, J. R. (1839). *Official report of the medical topography and climate of Calcutta with brief note of its prevalent diseases both endemic and epidemic*. G. H. Huttmann, Bengal Military Orphan Press.

Martin, M. (1836). Minute of the governor-general on the improvement of Calcutta, Fort William, June 16th 1803. In M. Martin (Ed.), *The dispatches, minutes and correspondence of the Marquess Wellesley KG during his administration in India* (Vol. IV, pp. 672–674). John Murray.

Ministry of Urban Development (MoUD). (2016a). *Municipal solid waste management manual, part 1: An overview*. Central Public Health and Environmental Engineering Organisation (CPHEEO), Ministry of Urban Development.

Ministry of Urban Development (MoUD). (2016b). *Municipal solid waste management manual, part III: The compendium.* Central Public Health and Environmental Engineering Organisation (CPHEEO), Ministry of Urban Development.

Mitra, M. (1990). *Calcutta in the 20th century: An urban disaster.* Asiatic Book Agency.

Mundoli, S., Manjunatha, B., & Nagendra, H. (2015). Effects of urbanization on the use of lakes as commons in the peri-urban interface of Bengaluru. *International Journal for Urban Sustainable Development, 7*(1), 89–108.

Mundoli, S., Unnikrishnan, H., & Nagendra, H. (2017). Nurturing urban commons for sustainable urbanisation in India. *Journal of the India International Centre, 43*(3–4), 258–270.

Nagendra, H. (2016). *Nature in the city: Bengaluru in the past, present and future.* Oxford University Press.

Narain, V., Banerjee, P., & Anand, P. (2014). *The shadow of urbanization: The peri-urban interface of five Indian cities in transition.* East West Center Working Papers No. 68, Hawaii, USA.

Narain, V., & Nischal, S. (2007). The peri-urban interface in Shahpur Khurd and Karnera, India. *Environment and Urbanization, 19*(1), 261–273.

Parkinson, J., & Tayler, K. (2003). Decentralized wastewater management in peri-urban areas in low-income countries. *Environment and Urbanization, 15*(1), 75–90.

Purushothaman, S., & Patil, S. (2019). *Agrarian change and urbanization in southern India: City and the peasant.* Springer Nature Pte.

Rees, W., & Wackernagel, M. (1996). Urban ecological footprints: Why cities cannot be sustainable—And why they are a key to sustainability. *Environmental Impact Assessment Review, 16*, 223–248.

Rooijen, S. J. V., Turral, H., & Biggs, T. W. (2005). Sponge city: Water balance of mega-city water use and wastewater use in Hyderabad, India. *Irrigation and Drainage, 54*, S81–S91.

Saldías, C., Speelman, S., Drechsel, P., & Huylenbroeck, G. V. (2017). A livelihood in a risky environment: Farmers' preferences for irrigation with wastewater in Hyderabad, India. *Ambio, 46*, 347–360.

Scott, C. A., Faruqui, N. I., & Raschid-Sally, L. (Eds.). (2004). *Wastewater use in irrigated agriculture: Confronting the livelihood and environmental realities.* CABI Publishing and Colombo: International Water Management Institute and Ontario: International Development Research Institute.

Shaw, A. (2005). Peri-urban interface of Indian cities growth, governance and local initiatives. *Economic and Political Weekly, 40*(2), 129–136.

Simon, D. (2008). Urban environments: Issues on the peri-urban fringe. *Annual Review of Environment and Resources, 33*, 167–185.

Simon, D., & Adam-Bradford, A. (2016). Archaeology and contemporary dynamics for more sustainable, resilient cities in the peri-urban interface. In B. Maheshwari, V. P. Singh, & B. Thoradeniya (Eds.), *Balanced urban development: Options and strategies for liveable cities* (pp. 57–84). Springer Open.

Smit, J., & Nasr, J. (1992). Urban agriculture for sustainable cities: Using wastes and idle land and water bodies as resources. *Environment and Urbanization, 4*(2), 141–152.

Thippaiah, P. (2009). *Vanishing lakes: A study of Bangalore city* (Social and economic change monographs no. 17). Institute for Social and Economic Change.

United Nations Development Programme (UNDP). (2015). *Transforming our world: The 2030 agenda for sustainable development.* United Nations Development Programme.

Unnikrishnan, H., Manjunatha, B., & Nagendra, H. (2016). Contested urban commons: Mapping the transition of a lake to a sports stadium in Bangalore. *International Journal of the Commons, 10*(1), 265–293.

Vij, S., Narain, V., Karpouzoglou, T., & Mishra, P. (2018). From the core to the periphery: Conflicts and cooperation over land and water in periurban Gurgaon, India. *Land Use Policy, 76*, 382–390.

Yadav, K. K., Singh, P. K., & Purohit, R. C. (2016). Impacts of wastewater reuse on peri-urban agriculture: Case study in Udaipur city, India. In B. Maheshwari, V. P. Singh, & B. Thoradeniya (Eds.), *Balanced urban development: Options and strategies for liveable cities* (pp. 329–339). Springer Open.

Chapter 3
From Royal Canal to Neglected Canal? Changing Use and Management of a Traditional Canal Irrigation System in Peri-Urban Kathmandu Valley

Anushiya Shrestha, Dik Roth, and Saroj Yakami

3.1   Introduction

While Nepal is the least urbanized country in South Asia, the current rate of urbanization in Kathmandu Valley, where the country's capital Kathmandu is located, is among the highest in this region (Muzzini & Aparicio, 2013). The valley population, growing at about 4.3% annually, increased by more than 499% between 1955 and 2008 (Bhattarai & Conway, 2010). It increased from 1.6 million in 2001 to more than 2.5 million in 2011 (Central Bureau of Statistics, 2001, 2012), and continues growing rapidly. This pace and intensity of urbanization and population growth cannot take place without seriously encroaching on the valley's land and water resources. Ongoing urbanization is causing land speculation, rising land prices, and radical land use changes, which also lead to a significant change in water uses and increase in water demand. Both quality and quantity of surface water and groundwater are under serious pressures, while the gap between water demand and supply continues growing (KUKL, 2017; Pandey et al., 2012; Shrestha et al., 2015).

A major aim of the research on which this chapter is based was to explore how these changes are experienced in specific water use settings in peri-urban Kathmandu Valley, and to assess their consequences for livelihoods, water rights, water access, water security, and the occurrence of water-related conflicts. Such insights can be gained from in-depth field research on major changes in use and management

A. Shrestha (✉)
South Asia Institute of Advanced Studies (SIAS), Kathmandu, Nepal

D. Roth
Sociology of Development and Change group, Wageningen University,
Wageningen, The Netherlands
e-mail: dik.roth@wur.nl

S. Yakami
Wageningen University, Wageningen, The Netherlands
e-mail: syakami@metameta.nl

© The Author(s) 2022
V. Narain, D. Roth (eds.), *Water Security, Conflict and Cooperation in
Peri-Urban South Asia*, https://doi.org/10.1007/978-3-030-79035-6_3

practices of key water sources and water infrastructure, in an overall governance context that is urban-biased. The stream-fed surface irrigation canals of Kathmandu Valley, with their long-established and—until recently—relatively stable water rights, provided a good setting for such research. Until urbanization of Kathmandu Valley took off in the 1980s, the management of stream-fed canal irrigation systems had been a priority of both state agencies and the population that depended on agriculture-based livelihoods. The name *rajkulo* (royal canal) given to these systems expresses the historical interests of (royal) state actors in canal maintenance and management.

Mahadev Khola Rajkulo is such a traditional canal irrigation system that has come under increasing pressure from urbanization. In this chapter we focus on developments around this canal, which is fed by a stream called Mahadev Khola in Dadhikot, a peri-urban village in Bhaktapur District in Kathmandu Valley.[1] This canal system is believed to have been developed around the 1670s (Adhikari, 2012). Initially run by villagers, it was managed by the Department of Irrigation from the 1950s until the early 1980s, and then handed over to a formally established water users' association (WUA) in the early 1990s. This WUA, however, did not remain active for long, with negative consequences for canal management. Meanwhile, changes like growing population, more intensive and increasingly commercial water uses, and decreasing reliability of the canal had driven those who could afford it into private investments for groundwater.

While struggles against water insecurity in agriculture and beyond are continuing, government policies have since long lost interest in action to keep peri-urban agriculture productive, and increasingly prioritize investments in urban expansion instead. We analyse interlinkages of the demographic, socio-environmental, economic and political-institutional dynamics with the changing uses and management of the canal. Further, we trace the actors' association with, or dissociation from canal uses and management through time, and what these processes mean for water access, water rights and water security. Our research shows that land use, social and institutional changes have led to a gradual weakening of canal-based water rights. Access to water from the canal is increasingly shaped by access to land in the upper reaches of the canal, to capital and technology, and to the right connections and social networks for funding and political support. These support new claims to water rights and new pragmatic practices of accessing water, rights-based or not.

This chapter is structured as follows: Sect. 3.2 provides an explanation of our research approach and methodology, and a short introduction of Kathmandu Valley and Dadhikot, the study area. In Sect. 3.3 we discuss the history of canal management and canal uses of the Mahadev Khola Rajkulo, with a focus on the last decades. In Sect. 3.4 we reflect on the implications of these changes in terms of changing water access, water rights, water security and water-related conflict, in a context of national urban policies that formally aim to conserve agricultural land and

---

[1] Administratively, Kathmandu Valley includes three districts: Kathmandu, Lalitpur and Bhaktapur. In the sixteenth century, Bhaktapur was the capital of the Kathmandu Valley Kingdom. The king constructed the irrigation canal to bring in water for drinking and irrigation (Pradhan, 2012).

livelihoods in this rapidly urbanizing valley but hardly contribute to their institutional support in practice. The chapter ends with a short conclusion highlighting the existing gaps in irrigation management in relation to the policy aim and practical need of conserving agricultural land.

3.2 Research Approach, Methodology and Study Area

3.2.1 Research Approach and Methodology

This study was inspired by the question how urbanization and associated changes in peri-urban spaces (see Simon, 2008; Leaf, 2011; see Chap. 1) influence the use, management and control of surface and groundwater in Kathmandu Valley. As shown by earlier studies, the rapid transformations caused by urbanization put enormous pressures on peri-urban spaces: growing settlements and population densities, migration and diversification of economic activities, emerging land markets, and changing land and water uses, to mention some. These often take place in increasingly complex socio-political and institutional environments (Butterworth et al., 2007; Simon, 2008; for South Asia, see Narain & Prakash, 2016). Recent changes in the use, management and governance of water in Kathmandu Valley provide a clear example of these processes. Urbanization and the expansion of water-based livelihoods bring in new water uses and users, and new ways of governing land and water, related to new development agendas and policy priorities. These are deeply influencing the, until some decades ago, relatively stable and ordered land and water access, rights, and use practices in the valley. They also lead to an ongoing erosion of local practices of negotiating water rights and solving water conflicts (see Upreti, 2000).

We approach these changes around the Mahadev Khola Rajkulo through a conceptual focus on water rights, approached from a legal anthropological property perspective (von Benda-Beckmann et al., 2006; see Roth et al., 2005, 2015) and access theory (Ribot & Peluso, 2003). In this approach, water rights are more than just the right to a share of water. A water right basically concerns "bundles" of rights and obligations pertaining to water and related resources (e.g. infrastructure) (von Benda-Beckmann et al., 2006; Roth et al., 2015). In canal irrigation, such bundles of rights and obligations have been conceptualized as "hydraulic property" (Coward, 1986; Coward & Levine, 1987). Following Ribot and Peluso's (2003) theorization of access—in relation to property—access is defined as "the ability to benefit" from material or immaterial resources such as water or social networks and relationships. As we will show, changing land and water uses, water rights and access, and organizational practices of canal management and maintenance of the Mahadev Khola Rajkulo are closely related. It is especially the intricate interplay between rights and access—not only to land and water, but also to funding, markets, infrastructure, technology etc.—in wider processes of socio-technical, demographic, economic and institutional change that deeply influences people's experiences of water (in-) security (Lankford et al., 2013; Zeitoun et al., 2013) and may also increase the probability of water-related conflicts (Shrestha et al., 2018).

As our aim was to understand how the use and management of an old irrigation canal system changed in the context of rapid urbanization and how this relates to the changing engagements of various actors and social relations between them, we designed our study as a qualitative case study of the Mahadev Khola Rajkulo. Case studies are well-suited for this type of research, in which in-depth information on the case forms the basis for a better understanding of the wider processes and relationships that the case represents, mainly related to transformations in water rights and access, and their implications for water security and conflict in a peri-urban context (see Flyvbjerg, 2006; Yin, 2009). The chapter is based on ethnographic research, with an interdisciplinary and socio-technical focus. The fieldwork for this chapter was conducted between 2015 and 2018 by the first and third authors of this chapter.

The findings discussed in this chapter are based on open-ended and conversational interviews, informal talks and group discussions with a wide range of informants. Interviewees included 26 farmers (18 male, 8 female), in-migrants (3), members of community-based organizations (7), executive committee members of the WUA (9), many of whom were either former or present local elected representatives. Interviews were also conducted with relevant government officials at the village level (1), municipal level (4) and district level (2), as well as from the department of irrigation (4). Changes in land and water use, rights and access to canal irrigation, and the roles and responsibilities in canal management were major topics during the interaction with farmers. These were also discussed with the executive members of the WUA and of community-based organizations, many of whom were also farmers. Interactions with these groups additionally helped to reflect on the formation, functions and dysfunctioning of the WUA and the alternative canal management groups and practices that have emerged over the years. Interviews with the government officials focused on the implications of peri-urbanization on food and water security, the state-led interventions for promoting agriculture and ensuring irrigation services in peri-urban settings, and challenges in these initiatives. In addition to these interviews, observation of the encroached canal sections, the intake, deserted field office and poorly maintained and leaking canal sections provided important insights in understanding the changing canal use and management practices.

3.2.2 *Kathmandu Valley, Dadhikot and Mahadev Khola Rajkulo*

Kathmandu Valley is one of the most productive agricultural areas in Nepal, with yields of major food grains (rice, wheat, maize) significantly higher than national averages (HMG & USAID, 1986). Historically, the valley exemplified the only autonomous urban system based on productive agriculture and long-distance trade in Nepal (Ministry of Urban Development, 2017). The earlier settlements in this valley were located on drier, less fertile, elevated land. The low-lying farming areas of Kathmandu Valley had an elaborate network of irrigation systems and were known for being highly productive (HMG & USAID, 1986; Pradhan & Belbase, 2018). This land use practice was based on a conscious land use strategy, intended to maximize the area of land used for agricultural purposes and to preserve fertile and irrigable agricultural land (Shrestha & Shrestha, 2009; Tiwari, 1999).

The early urban growth of Kathmandu was based on its agricultural surplus (Ministry of Urban Development, 2017). Joshi (2018) notes that the rich cultural tradition of Kathmandu Valley can be attributed to the network of irrigation systems that supported advanced and intensive agricultural activities which, in turn, sustained a highly developed urban culture. Irrigation in Kathmandu Valley was possible through stream-fed canal irrigation systems. The operation and management of many irrigation canals received royal state sanction and support (Dixit, 1997). Hence these canals were called *rajkulos*, "royal canals". Originating in the foothills, *rajkulos* are long-distance water canals, mostly built in the seventeenth century to fill up ponds, irrigate farmland along the canals, and serve settlements and their agricultural areas in the valley (GoN/NTNC, 2009; Shrestha & Shrestha, 2009). Irrigated agriculture was both a spiritual and material foundation of community life and was coupled with the *Guthi* system, a traditional socio-cultural and religious institution (Joshi, 2018).[2]

After the Department of Irrigation had been established in 1952, many farmer-managed irrigation systems in Kathmandu Valley were rehabilitated (Pradhan & Belbase, 2018). In the same period, urban growth outside the historic city core into the valley also started. This urban expansion gained momentum in the 1970s, as more areas were made accessible with the construction of a ring road, while establishment of industries increased employment opportunities. Urban expansion accelerated in the 1980s and 1990s, deeply transforming the rural agricultural landscape of the valley into peri-urban spaces characterized by a co-existence of agricultural and non-agricultural land and water uses, economic activities and livelihood practices.

The growth of settlements in the valley is generally spontaneous, as there is hardly any land use and other planning by government agencies (International Center for Integrated Mountain Development, 2007).[3] This trend of unplanned urbanization has continued, engulfing peri-urban and rural agricultural land at an alarming rate (Ministry of Urban Development, 2017). However, although the percentage of land used for agriculture has drastically declined, thanks to its unique soil formation and topographical variation the agricultural potential of Kathmandu Valley remains considerable even today (NARC, 2006, cited in Rana et al., 2015). The Ministry of Urban Development acknowledges the need for promotion of urban

---

[2] *Guthi* is a traditional institution for collective, mainly religious and cultural, activities.

[3] Implementation of various plans for urban development of Kathmandu Valley has been poor (Ministry of Urban Development/KVDA, 2015).

and peri-urban agriculture "in rooftops, vacant plots, peripheral areas of public buildings, parks, garbage landfills, etc." to ensure food security now and in the future (Ministry of Urban Development, 2017: 34).[4] Providing about 33% of the gross domestic product (GDP) and supporting the livelihoods of most of the population, agriculture is a mainstay of the Nepalese economy. Irrigation is crucial in maintaining and increasing agricultural production (Gautam, 2012; Pradhan & Belbase, 2018).

With an area of 6.27 km², Dadhikot[5] is a rapidly growing peri-urban space, around 12 km east of Kathmandu. Dadhikot became part of Anantalingeswor Municipality,[6] which merged with Suryabinayak Municipality in a restructuring of the local government in 2017 (Fig. 3.1). The annual population growth of Dadhikot

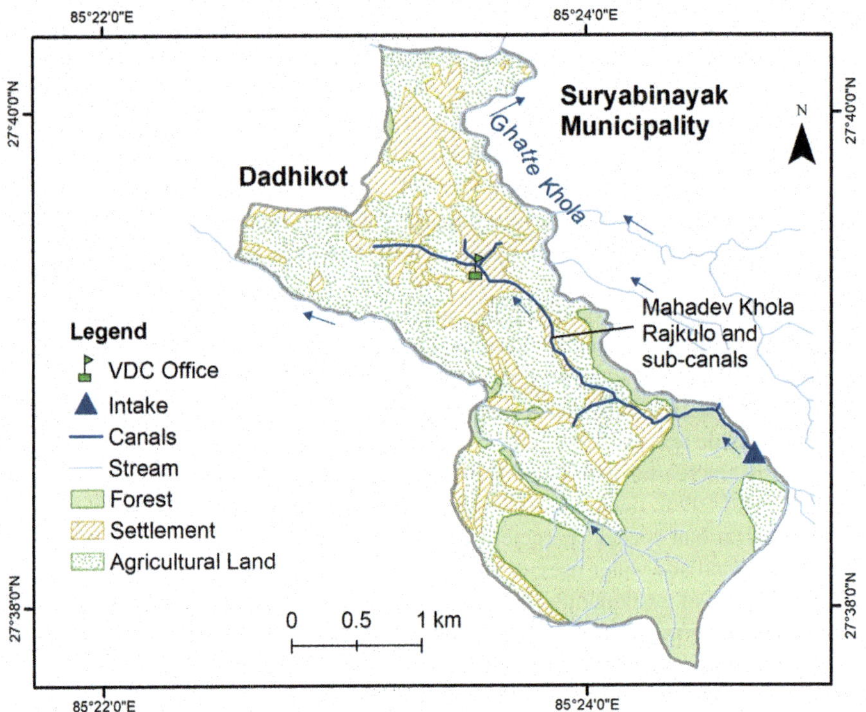

Fig. 3.1 A section of Mahadev Khola Rajkulo and sub-canals in Dadhikot

[4] Kathmandu Valley produces 4.6% of the vegetables and 3.5% of the potatoes produced in the country (Ministry of Agriculture, 2012 cited in Ministry of Urban Development, 2017).

[5] Until 2014 Dadhikot was a Village Development Committee (VDC). A VDC referred to a rural administrative unit and used to be the smallest local government unit in Nepal. Each VDC was administratively divided into 9 wards.

[6] The name "Anantalingeswor" comes from a temple believed to originate in the seventh century. With connections to the then ruling family, the area had administrative and political significance (see Shrestha, 2007).

increased from 1.17% (1981–1991) to over 6% by 2011. The built-up area has increased by over 250% between 1992 and 2010, and is expected to grow by about 110% between 2010 and 2030 (Sada et al., 2016). Nonetheless, agriculture still provides a major livelihood for many of its inhabitants. Paddy is the most common monsoon crop. In winter, many farmers have switched from wheat farming to commercial vegetable farming. Agriculture still depends on the traditional surface canal irrigation systems fed by the Mahadev Khola, a stream in Dadhikot. Mahadev Khola Rajkulo is the largest canal irrigation system in Dadhikot. Because of its reliable water flow in the past, many water mills for grinding wheat used to be operated along the canal; hence it was also called *Ghatte Kulo* (canal with water mills).

3.3 From Glorious Past to Neglected Present: The Trajectory of Mahadev Khola Rajkulo Management

3.3.1 *The Glorious Past of the Mahadev Khola Rajkulo*

The Mahadev Khola Rajkulo is believed to be have been developed in the 1670s by the then king (Adhikari, 2012). Every year, farmers made a temporary intake of stones, brushwood and clay, through which water entered the earthen canal. The system was strongly associated with the socio-cultural and religious values of the kingdom. According to local farmers, a large area of land irrigated by this *rajkulo* was under *guthi* tenure.[7] In a historical study on the Anantalingeswor area, Shrestha (2007) mentions that this ancient *guthi*, termed *"Ganesh Rajguthi"*, had royal significance and was managed by an ethnic group of Dadhikot, *Rajthala*. These Rajthala were relatives of the royal family of ancient Bhaktapur. With a significant political-administrative role and surrounded by highly productive land, this was a prosperous settlement.[8] Through its control of *guthi* land, *Rajthala* was the primary group involved in the management of the canal. They cultivated the *guthi*-land, and part of the harvest was allocated for canal management. A farmer:

> The harvest was annually collected in front of the temple in our village. My grandfather told me that the pile of harvest collected used to be higher than the temple, which had two storeys then.[9]

This harvest share was used, among others, to mobilize manpower for canal repair and maintenance. As the canal was of prime importance for agriculture-based livelihoods, the *guthi* institution organized rituals at a local temple, which is believed to

---

[7] *Guthi* land refers to the land donated by the state or individuals for the purpose of religious or charitable institutions. *Guthi* land was exempted from tax and could not be taken back by the donors. Hence, people had a preference for *guthi* land. This community land tenure system continued even after the unification of Nepal in 1769 (Adhikari, 2008).

[8] Earlier known as Chawadesh, this settlement is called Chitrapur (see Shrestha, 2007).

[9] Before the earthquake in the early 1930s damaged it.

be the source of water for the Mahadev Khola, and thus the canal. In years with a delayed monsoon, special rituals for requesting rain were held.

As the earlier prosperity and political role of this settlement declined with changes in the political regime (Shrestha, 2007), the cultural practices associated with the *rajkulo* started decaying in the 1960s. A key informant: "during the inventory of landownership in the mid-1960s, many farmers claimed to be owners of the land [they worked] and thus converted *guthi* land into private land. After a few years, *guthi* and related practices were discontinued". Yet, as canal irrigation continued to be crucial for agriculture, farmers remained widely involved in canal cleaning and maintenance. This practice of management by the farmers may also have been driven by the *Muluki Ain*[10] of 1854, the first codified and written law applicable to the whole kingdom, and later amendments to it in 1952 and 1963 (Pradhan, 2000). These later versions allocated canal maintenance responsibility to tenant farmers, and even made their tenancy rights dependent on their labor and other contributions (ibid). If tenant farmers were unable to repair the canal system, the landowners often requested the government for a grant. Our informants stated that, until enactment of the Land Act of 1964, which entitled tenant farmers to half of the cultivated land, land in Dadhikot was largely owned by landowners from an adjoining settlement (Madhyapur Thimi Municipality). Farmers from Dadhikot were mostly tenants, and thus responsible for maintenance of the canal, the life-line of their agriculture-based livelihood.

3.3.2   Government Intervention in Irrigation and Re-institutionalization of Community-Led Irrigation Management

According to Pradhan (2000), state control of water resources has rapidly increased since the 1950s, especially since promulgation of the 1992 Water Resources Act.[11] Before the 1950s most regulation (in the *Muluki Ain*) focused on land tenure; while land was seen as a major revenue source, water and its regulation were seen as unimportant and left to local customary laws and uses. This changed after 1951, when public sector involvement in irrigation development started. In the 1950s, foreign aid also started entering the country, bringing about further socio-economic and legal changes. During these initial years of government intervention in irrigation, the Irrigation Department established individual offices for specific projects and took monitoring and maintenance responsibilities for completed projects (Pradhan & Belbase, 2018). The permanent dam of the Mahadev Khola Rajkulo

---

[10]The *Muluki Ain* has been adapted many times, and was recently replaced by the Civil and Criminal Codes of 2018.

[11]A 'revolution' in the crucial 1950–1951 period ended the Rana autocratic regime and established parliamentary democracy under a constitutional monarchy (Pradhan, 2000).

Fig. 3.2 Dam of Mahadev Khola Rajkulo. (Photo Anushiya Shrestha)

(Fig. 3.2) was constructed in 1956 (District Irrigation Office Bhaktapur, 1996), as a result of which the command area of the canal increased.[12] Two caretakers[13] and a supervisor, with a field office near the intake, were appointed under the responsible department for monitoring and operation of the dam and canal. They were involved in the operation of the barrage gates, canal maintenance and water tax collection.

In the late 1980s, the government introduced its Irrigation Working Policy, which emphasized participatory planning, development and management of irrigation schemes (Pradhan & Belbase, 2018).[14] This new planning approach required a formally registered WUA for the construction, operation and maintenance of irrigation systems. In 1990, the multi-party system was declared. According to the farmers, the new government under the multi-party system eliminated the position of supervisor and reduced the number of appointed caretakers for Mahadev Khola Rajkulo to one. The latter provision was also slashed after 1992. In the same year, a landslide damaged the main canal at its head reach, which resulted in the complete dysfunctioning of the irrigation system for some years. As part of efforts to reconstruct it, a

---

[12] The new gross command area was 625 hectares and the net command area 450 hectare (District Irrigation Office, 1996), including Dadhikot and a downstream VDC.

[13] The caretakers were selected among the local farmers so that the operation, repair and maintenance of the main and the branch canals could be performed timely and through mobilization of the local farmers.

[14] The Irrigation Regulation and Irrigation Directives of 1988 were also brought out by the government in sequence. The Irrigation Directives provided detailed procedures for the formation of WUAs (Pradhan & Belbase, 2018).

WUA with members from Dadhikot and the downstream VDC irrigated by this Rajkulo was formed around 1994. A key-informant recalls:

> after the massive landslide in 1992 we were not able to operate the canal for some years. We bitterly needed irrigation. Without irrigation, production would decrease, which could cause social problems like theft. So canal maintenance was essential. Finally, we agreed to use a temporary pipe to operate the canal. We formed a WUA and obtained NPR 800,000 from the government for the maintenance".[15] [....] Although the WUA had to actively engage in and contribute to the repair work, it was done through a contractor. A mistake was made in repairing the section, which further damaged the *rajkulo*.

This started disputes between the WUA members, with one member accusing another for intentionally causing this damage with the purpose of misusing the grant. Members of the WUA recall that, although the feasibility study of the irrigation project had been conducted, the work could not start due to the approaching local election in 1996. The newly elected local representatives claimed that they had the right to be WUA members and demanded reform of the recently established WUA. After much debate, the WUA was reformed, replacing some sitting members by elected local representatives. Debates about inclusion in and exclusion from the WUA created further internal divisions. The rehabilitation was completed amid increasing disagreements on political interests, financial mismanagement and lack of transparency related to the rehabilitation contract. Although the functioning of the canal had improved after rehabilitation, the continuing disagreements ruined the relations among WUA members; hence the WUA turned inactive soon after.

3.3.3 Changing Water Allocation and Use in the Context of Increasing Urbanization

Although the rehabilitation had improved irrigation services, covering the growing drinking water needs from a few traditional sources remained a daily hardship in Dadhikot. To solve this, the then elected local representatives of Dadhikot initiated a community-managed drinking water supply system in the mid-1990s. This system, based on a local spring source, started regular services via public taps, and shifted to household-connections with increasing urbanization and growing water demands. In both national and donor policies, improved drinking water (and sanitation) services were increasingly prioritized.[16] Many other smaller community-level drinking water services based on local spring sources were initiated or rehabilitated soon, largely with government support. Many members of the Mahadev Khola Rajkulo WUA

---

[15] This financial assistance was under the government-supported Second Irrigation Sector Project (SISP) (District Irrigation Office, 1996), financed by the Asian Development Bank. Farmers had to contribute in cash and labor in accordance with the participatory management approach.

[16] The Water Resources Act (WRA) of 1992 prioritized access to drinking and domestic water over irrigation. Internationally, 1980-90 was the International Drinking Water Supply and Sanitation Decade (see Castro, 2007).

were involved as functionaries of such systems. As they started competing with each other to expand their services, an attempt to revive the WUA around 2005 failed. In the meantime, the available institutional and financial support and rapidly increasing water demands by a growing population provided further incentives for the local government and community-level water managers to expand drinking water supply services. As such initiatives competed with agricultural water use, they were seen as a potential threat by farmers of irrigated crops. According to one of them, "in the mid-2000s, efforts were made by a drinking water supply institution to tap water at the intake of the irrigation system. Farmers strongly protested this".

Although farmers were able to resist this attempt to appropriate their water source, availability of canal water for Mahadev Khola was declining from the mid-1990s, when a permanent dam was built in the upstream village (where the Mahadev Khola originates). Traditionally, upstream villagers could only block water using brushwood (not stones), a method referred to as *syaule bandh* (brushwood dam). This practice guaranteed a sufficient water flow downstream. Irrigation management regulations restricted any construction that would reduce the share of water for downstream farmers. Earlier upstream attempts to make a permanent dam had been strongly resisted by Dadhikot farmers and the downstream village depending on Mahadev Khola Rajkulo.

However, while these downstream farmers were struggling with the consequences of the landslide that obstructed their irrigation, in the mid-1990s upstream villagers constructed their permanent dam. According to Dadhikot farmers, this was possible through the political affiliation between the chairpersons of Dadhikot and the upstream VDC. The share of water for this canal system, which was reduced by this construction, further decreased over the years with an increasing use of spring sources for drinking water supply, degradation and drying up of traditional ponds, and declining rainfall.[17] With urban expansion, the commercial value of land and the number of land transactions have rapidly increased, changing land and water uses and livelihoods in this agriculture-based village. Poor canal maintenance, declining canal water availability and increasing upstream-downstream competition for scarce canal water further reduced farmer involvement in canal maintenance. Under the Local-Self Governance Act (LSGA) of 1999, the irrigation system became a responsibility of the local government as the custodian of natural resources and public infrastructure.[18] At the request of the farmers, the village government provided a small annual budget for dam operation during the monsoon season and some assistance for canal cleaning and maintenance. Meanwhile, the district irrigation offices in Kathmandu Valley had been replaced by an irrigation division office. More importantly, with changes in irrigation policies, government assistance for irrigation had changed into a demand-driven approach, requiring an active WUA—which was absent in Mahadev Khola Rajkulo.

---

[17] Average annual temperature in Kathmandu Valley shows an increasing trend (0.033 °C/year), while total annual rainfall shows a decreasing trend (−5.9 mm/year) (UN-Habitat, 2015).

[18] The Local Government Operation Act of 2017 has been formulated under the federal system of governance.

3.3.4   Urbanization, Declining Irrigation Service and Changing Priorities

With an inactive WUA and decreasing water availability for irrigation, the water sharing arrangement between Dadhikot and the downstream VDC was discontinued.[19] Furthermore, with a decreasing flow and increasing canal leakage, access to irrigation water for Dadhikot farmers has also declined. Year-round irrigation service nowadays has become restricted to farmers with fields near the canal intake; for the others it has become limited to the monsoon season. Even head reach farmers need to compete for a timely and equal volume of water, sometimes resulting in disputes. These escalated with the switch of winter crop from wheat for subsistence to commercial vegetable crops, an important new source of income here. This requires a regular irrigation of low intensity, increasing water competition between upstream and downstream farmers and sometimes causing conflicts. According to a farmer:

> Upstream farmers grow pumpkin and cucumber for commercial supply. They only harvest these towards mid-July. If [too much] water flows into their fields, their crops will be damaged so they do not let the water go into the fields until harvesting is completed. As our fields depend on a branch canal fed with overland flow from these upstream fields, we do not get water even if it is in the canal. After harvesting, these upstream farmers are already in a rush for transplanting paddy and competing among themselves for canal water to prepare their fields. So there is little chance that we [lower-reach farmers] get canal water [even then].

Yet, farmers opine that canal water-related conflicts are fewer than in the past. A farmer from the middle-reach area:

> Although not violent, minor quarrels were common in the past, particularly in the monsoon during the paddy transplanting period. In recent years, this has drastically decreased. Farmers in the middle and lower canal reaches know that upper-reach farmers will not leave water for the lower-reach farmers until they have completed their own irrigation.

This seeming contradiction—less water, but also less conflicts about water—may be explained by the fact that the decreasing reliability of the canal has set in motion a trend towards investments in technology for individualized water provision. Many farmers in the lower reaches have sunk dugwells or borewells and use groundwater for irrigation. However, lower-reach farmers stress that groundwater is not sufficient for irrigation, as they also use it as their alternative for declining drinking and domestic water services.

Yet, except for a few wealthy commercial farmers who have individually installed drip irrigation fed by their groundwater source, most farmers have to make do with limited groundwater or depend on rainfall. Thus, managerial concern for the canal continues decreasing. A former elected local representative and former WUA member:

---

[19] Farmers in Dadhikot irrigated during day-time (5 am to 5 pm), downstream village farmers during the night.

> Policy is driven by the motive of urban expansion. With expanding built-up areas, people are increasingly discharging their household sewerage into the canal. We have grown up using drinking water from the canal, but now we hesitate even to put our foot in the canal. In a meeting at the municipality, I proposed to keep some hoarding boards to warn against this and stressed the need to improve canal management, but nothing has been done. It seems that they assume that [as in many other municipalities] this is a canal for now and will be a sewer in the future.

Meanwhile, some initiatives for canal maintenance have been taken with government funds. Female farmers, engaged in commercial farming through their women groups,[20] are maintaining small sections of a branch canal of Mahadev Khola Rajkulo through a small irrigation project under the District Agriculture Development Office (DADO). However, such activities are limited, while the thin and low-quality concrete canal lining gets quickly damaged, particularly as water flows increase during the monsoon season. While such groups are often blamed for bad repair works, women farmers' group members explain that they are required to maintain a given length of canal section and that beneficiary farmers should make a significant financial contribution to such works. However, as the farmers do not contribute, they are compelled to make do with the limited government funds, as a consequence of which the repair work cannot sustain. Other farmers argue that poor monitoring by DADO allows women farmers' group members to misuse funds. The DADO office, however, has limited human resources at the field level and lacks technical expertise regarding physical infrastructure.

Despite extensive monitoring, the maintenance supported by the district government did not sustain either. A farmer:

> In 2016, when district government support was available for the canal, an ad-hoc committee was formed. Its chairperson has good political links. A check dam was constructed to protect the dam against [too strong] water flows in the monsoon. This was done under contract. In the same year, after a few days of monsoon showers, the check dam was washed away. The damage caused dissatisfaction among the farmers. Yet, in the next year the contract was again given to the same person.

Farmers tell that, with negligence in monitoring and evaluation, canal maintenance activities have actually not benefited the farmers, as the latter continue to suffer from canal leakage. In a group discussion on canal-related issues, women farmers stated:

> We cleaned the canal and tried to bring water to our fields. But due to the leakage and seepage along the canal, water hardly reaches our fields. If this canal can be improved, we will have adequate water for irrigation.

Realizing the importance of repairing the dam, the canal and its operation, in 2016 the head reach farmers submitted a request at the Irrigation division office (IDO). The feasibility study conducted following this application showed that canal repair and maintenance was economically viable. However, the project was not selected

---

[20] Members of farmers' groups are entitled to various services, like training and input materials for agriculture.

for implementation. In response to queries about the reasons for this, a high-level official from the Department of Irrigation (DoI) argued:

> Peri-urban areas are rapidly urbanizing, so investing in irrigation for staple food is useless. We can go for drip and sprinkler for cash crops, if there is demand and farmers are willing to contribute to the investment. Even for that, water availability is crucial. For peri-urban areas, the government will not invest if investments are high and return is low.

In addition, with the transfer towards the municipality status, expansion of roads and the Transportation Masterplan have become the focus, upgrading foot trails into motorable roads.[21] Unlike for irrigation, as pointed out by the official, the economic return of investments in urban infrastructure (water supply and sanitation, solid waste management, and transportation) is expected to be substantial through direct income, capital gains and increases in taxes from urban residents (Ministry of Urban Development, 2017).

Investing for irrigation services in peri-urban areas is clearly low. As elaborated above, for many farmers lack of reliable irrigation services has been a major impediment to continuation and expansion of their agriculture-based livelihood. Further lack of irrigation may reduce the effectiveness of government activities to promote agriculture in urbanizing areas. An official from the District Agriculture Development Office explained:

> Agricultural lands are being converted into residential plots and sold. Likewise, there are a large number of brick kilns under operation. Solving all these problems is beyond our capacity but we are promoting [peri-urban] agriculture and advocating for the same. Land in this district is highly productive and we envision developing 'model agro-farms'. While we [Agriculture Department] only have small support for irrigation, our programs such as subsidy on seeds, fertilizers, plastic sheets, tractors have direct benefits for the farmers. We also have programs to attract the youths in agriculture to generate self-employment opportunities and control their outmigration in lack of employment. These programs are under the support of the central government. This year we are promoting collection centers and have market extension programs. Until agriculture is commercialized, neither agriculture nor the country can be developed.

Urbanization has created new markets and thus stimulated commercialization of agriculture in peri-urban Kathmandu Valley. Examples are the switch towards commercial farming, particularly vegetables and floriculture that are commonly seen in areas with access to reliable irrigation. With urbanization, the commercial value of land has also skyrocketed, which has tempted many farmers to sell land (Shrestha, 2011). In Dadhikot, rising land prices and the widening of roads in an overall context of increasing land fragmentation,[22] declining areas of holdings, worsening irrigation services and, consequently, decreasing crop yields, are creating an enormous incentive for farmers to sell their land. With the government-planned construction of an outer ring road passing through Dadhikot, urbanization and associated socio-environmental transformations will continue and probably intensify, while the canal and those depending on it for irrigation will lose in importance.

---

[21] About 60% of the costs in existing urban areas is earmarked for upgrading and extension of roads (MoUD, 2017).

[22] As a consequence of division of land between tiller and landowner, and inheritance practices.

3.4 Changing Canal Use, Rights and Access to Water, and Water Conflict

In this chapter we have shown how the Mahadev Khola Rajkulo in Dadhikot has undergone many changes in its long history, and how and why in these processes various actors associated with, and dissociated from, its use and management. Big changes in the more or less stable "traditional" socio-cultural practices of canal management started in the 1960s, and intensified under the influence of development programs from the 1980s. But the most radical changes happened since the 1990s, with the increasing pace of urbanization of Kathmandu Valley and changing developmental policies and priorities. Construction of a permanent dam in the upstream section was an outright, politically supported, denial of the existing customary water rights of farmers depending on this canal. As urbanization-driven changes in land and water uses continued, these farmers gradually lost their formerly strong and widely recognized rights and access to canal water in ever more complex ways (Fig. 3.3).

With changes towards participatory irrigation management policy, allocating management responsibilities to farmers and assuming their active participation by forming WUAs, the government has limited its own role as facilitator of irrigation systems management. In peri-urban contexts like Dadhikot, with rapidly changing land uses and increasing water competition, interests and participation in irrigation

Fig. 3.3 Diverse and changing uses of peri-urban land and water (These areas have largely turned rain-fed; see the dugwell in the middle of the field). (Photo Anushiya Shrestha)

management have been declining. Although a WUA was formed for the Mahadev Khola Rajkulo, its activities were limited to rehabilitating the irrigation system with external financial support. Similarly, activities of the local government, farmers' groups and ad-hoc committees for the canal seem to be a mere mobilization of development funds, with no significant long-term improvement of infrastructures and strengthening of institutional arrangements. Driven by the need to improve the canal system, upstream farmers attempted to form a new WUA. It could not materialize, however, because the proposed maintenance project was not selected for implementation. Over the years, irrigation services have become restricted to farmers close to the intake. Furthermore, with declining water availability and the switch to commercial farming, competition for canal water has spread to the riparian farmers in the upper reach as well.

Although open conflicts about canal water, which were common in the past, are not common anymore, divisions between upper-reach farmers with access to canal irrigation and an opportunity for irrigation-based economic activities, and lower-reach farmers without such access and opportunity, are prominent. Interestingly, during the fieldwork we noted that outsiders have also leased in land for commercial farming and even brick-making in the upper reach. They were able to use canal water by virtue of the water rights of the landowners (from whom they lease-in land) and by maintaining good social relations with neighboring farmers.[23] Those lower-reach farmers who could afford it created water access by investing in groundwater sources and other technologies (e.g. pumping stream water, drip irrigation system). Those unable to afford such alternatives depend on rainwater which, however, is increasingly unreliable. On the one hand, institutions, infrastructure and public interest in irrigation development are weakening, and thus adversely affecting agriculture-based livelihoods. On the other hand, the commercial value of land is increasing, which tempts farmers to sell their land. Urban expansion and in-migration have made this peri-urban village grow rapidly. Demand for drinking (and domestic) water and reallocation of water sources that earlier helped in irrigation have increased.[24] An attempt to reallocate the Mahadev Khola Rajkulo was strongly opposed by many farmers, and a similar attempt made in 2014 was supported almost exclusively by farmers from the upper reach.

The decline in management involvement of farmers from the lower reaches can be related to their worsening access to canal water, which has severely reduced their concern for the irrigation system. As has been illustrated above, interest in irrigation is not only decreasing among an increasingly diverse peri-urban population and at the local government level, but also in the government agencies responsible for irrigation development. With ongoing urbanization in the context of a changing

---

[23] We noted a brick-maker who had leased in land and hired a local inhabitant as a guard, whose responsibility included arranging water for brick-making by lifting water from the stream and even from the canal, when available.

[24] The Government of Nepal has the national target of universal access to safe drinking water and sanitation for all. According to the census data of 2011, 85% of the households have access to water supply and 61% have sanitation (Central Bureau of Statistics, 2014).

climate,[25] irrigation is likely to be an increasingly competitive and costly service, beyond the reach of peri-urban farmers, particularly those depending on canal irrigation. With declining irrigated agriculture-based economic activities, conflicts about irrigation water may no longer be an issue, but these changes in land and water uses will definitely have adverse effects on the formal goals of conservation of agricultural land, sustainable agriculture and food security.

The literature on property and access discussed above (Ribot & Peluso, 2003; von Benda-Beckmann et al., 2006) is particularly relevant to make sense of the changes around the canal. As Ribot and Peluso (2003) have stressed, the concept of access—the ability to benefit—is much broader than property. While a property right may create access to a resource like water, this is not guaranteed, as is well illustrated by the fate of the water rights that used to be strongly attached to the Mahadev Khola Rajkulo. Nor is access necessarily rights-based. Developments around the canal in the last decades are causing a gradual weakening of the "bundles" of property rights and obligations historically associated with the irrigation canal. While in earlier times rights-holding farmers could be relatively sure about their access to canal water, nowadays the situation has completely changed. Access to water in this urbanizing village is increasingly shaped by access to land in the upper reach, to capital and technology, and to the right connections and social networks for funding, bribing and political support. The many demographic, land use, social and institutional changes around the canal have led to a weakening of canal maintenance and management practices, while growing demands for water outside irrigated agriculture support new claims to water rights and new practices of accessing water, rights-based or not. These changes are deeply influencing people's experiences of water (in-)security (Lankford et al., 2013; Zeitoun et al., 2013) and may also increase the occurrence of water-related conflicts (Shrestha et al., 2018).

3.5 Conclusion

Dadhikot, like many other urbanizing villages in Kathmandu Valley, is undergoing rapid and radical changes, with great impact on its formerly agricultural land and water uses. A rapidly increasing population, unplanned urban development and loss of agricultural land in Kathmandu Valley have been identified as major peri-urban problems since long back, among others in the formulation of the first physical development plan of Kathmandu Valley developed in 1969 (His Majesty's Government of Nepal, 1969; Kathmandu Valley Development Authority, 2015). Several of the current national policy documents, including the urban policy of 2007, the urban development strategy of 2017, and the national land use policy of 2015, have stressed the need to control the rapid conversion of agricultural land into non-agricultural uses. However, the importance of reliable irrigation services in

---

[25] Climate change is noted to have adverse impacts on water resources in Nepal (WECS, 2011).

socio-economically sustainable agricultural livelihoods, and thus the need to restrict non-agricultural uses of land, has received little attention.

An example is the 20-years' Strategic Development Master Plan (2015–2035), drafted for the long-term development plan for Kathmandu Valley. While it recognizes the importance of food security and the need to protect "prime agriculture land" in the valley, it is silent about irrigation and lacks coordination with the organizations responsible for irrigation development (see Kathmandu Valley Development Authority, 2015). Moreover, as discussed above, irrigation management in peri-urban contexts is no longer a priority of the responsible authority, despite the fact that the irrigation policy recognizes prior rights over irrigation water and aims at conserving irrigated areas by restricting non-agricultural uses. As we saw in this study of changing use and management of a historic canal irrigation system in Dadhikot, the current practices with respect to water rights for irrigation contrast sharply with these policy intentions. In the meantime, the conversion of irrigated and other agricultural land into non-agricultural uses is continuing at high pace.

We share with Rana et al. (2015) the view that development plans and policies for Kathmandu Valley have remained biased towards the expansion of urban infrastructure. Despite its agricultural potential and importance, peri-urban agriculture lacks government support and recognition. Our study shows how the pressures on peri-urban agriculture and agriculture-based livelihoods are increasing, due to the low priority given to irrigation management. It is important to acknowledge that irrigation management is not a one-dimensional activity but rather a domain of complexly interconnected social, cultural, institutional, technical and agronomic activities (see Pradhan, 2000; Roth & Vincent, 2013). However, while irrigation management responsibilities have been allocated to communities, the local dynamics of changing land and water uses and how these shape community needs, interests and priorities, water rights and access to water, and conservation of agricultural land have received little attention.

Studies have stressed that water security is a relational, political and multiple-scale issue of both water access and control, and that any form of sustainable water security policy must consider such water-society interrelations and interdependencies with other resources (see Zeitoun et al., 2016). In the current national context of increasing priority for urbanization and ongoing efforts to strengthen national water policies, it is important to acknowledge that competition for water in peri-urban spaces is likely to increase with urbanization and the adverse impacts of climate change (WECS, 2011). Addressing climate change has been a major agenda in the formulation of new policies and strategies in Nepal, such as urban, irrigation, agriculture and water supply and sanitation policies. Through its National Adaptation Programme of Action (NAPA), The Climate Change Policy of 2011 has prioritized community-based adaptation though integrated management of agricultural, water and forest-based resources, to minimize the adverse impacts of climate change.

However, as discussed above, the rapidly changing land and water use practices and needs and interests of peri-urban communities largely contradict such formal community-based integrated management approaches. Studies on agriculture in peri-urban Kathmandu Valley show that this is a profitable endeavor (Bhatta & Doppler, 2016; Rai et al., 2019). Thus, interventions for improving irrigation management, institutions and infrastructures could stimulate the peri-urban population to continue agriculture, and thus prevent ongoing conversion of agricultural land into non-agricultural uses while dealing with existing water-related issues.

This is not to conclude that the lack of irrigation services is the main driver making farmers move away from agriculture, nor is urbanization the sole cause of the drying up of water sources or degradation of the irrigation canal system. Clearly, the multiple causes of the changes in peri-urban land and water uses and their consequences for water rights, access to irrigation services and agriculture-based livelihoods are complexly interlinked. Nevertheless, ignoring these rapid and rampant changes and the differentiated impacts these changes have for a socially and economically diverse peri-urban population can further intensify the conversion from agricultural to non-agricultural land (and water) uses, against the governmental policy of conserving agricultural land. Alternatively, supporting farmers in their efforts to continue agriculture by, for instance, improving irrigation services can help attracting them to agriculture while at the same time generating self-employment opportunities and avoiding adverse repercussions on food security.

References

Adhikari, J. (2008). *Land reform in Nepal: Problems and prospects*. Nepal Institute of Development Studies and Action Aid Nepal.

Adhikari, K. (2012). *Losing ground of collective action: Case study on operation and management challenges in Mahadev Khola Rajkulo emerging from urbanization and livelihood changes*. MSc thesis. Pokhara University. Changunarayan, Bhaktapur, Nepal: Nepal Engineering College.

von Benda-Beckmann, F., von Benda-Beckmann, K., & Wiber, M. (Eds.). (2006). *Changing properties of property*. Berghahn.

Bhatta, G., & Doppler, W. (2016). Smallholder peri-urban organic farming in Nepal: A comparative analysis of farming systems. *Journal of Agriculture, Food Systems, and Community Development, 1*(3), 163–180.

Bhattarai, K., & Conway, D. (2010). Urban vulnerabilities in the Kathmandu Valley, Nepal: Visualizations of human/hazard interactions. *Journal of Geographic Information System, 2*, 63–84.

Butterworth, J., Ducrot, R., Faysse, N., & Janakarajan, S. (Eds.). (2007). *Peri-urban water conflicts: Supporting dialogue and negotiation* (Technical paper series: No. 50). Delft, The Netherlands.

Castro, J. E. (2007). Water governance in the twentieth-first century. *Ambient Society, 10*(2), 97–118.

Central Bureau of Statistics (CBS). (2001). *Statistical year book of Nepal*. HMG.

Central Bureau of Statistics (CBS). (2012). *National population and housing census (Village development committee/municipality)*. Government of Nepal.

Central Bureau of Statistics (CBS). (2014). *Population monograph of Nepal. Volume II (social demography)*. Government of Nepal.

Coward, E.W. Jr. (1986). State and locality in Asian irrigation development: The property factor. In K.C. Nobe and R.K. Sampath (Eds.), *Irrigation management in developing countries: Current issues and approaches* (pp 491-508). : Westview Press.

Coward, E. W., & Levine, G. (1987). Studies of farmer managed irrigation systems: Ten years of cumulative knowledge and changing research priorities. In International Irrigation Management Institute (IIMI) and Water and Energy Commission Secretariat (WECS) of the Ministry of Water Resources, Government of Nepal (Ed.), *Public intervention in farmer-managed irrigation systems*. International irrigation Management Institute.

District Irrigation Office (DIO). (1996). *Feasibility assessment report of Mahadev Khola rajkulo sub-project. Ministry of Water Resources, Department of Irrigation, Central regional Irrigation Directorate, Second Irrigation Sector Project*. District Irrigation Office.

Dixit, A. (1997). Inter-sectoral water allocation: A case study in Upper Bagmati Basin. In R. Pradhan, F. von Benda-Beckmann, K. von Benda-Beckmann, H. L. J. Spiertz, S. Khadka, & K. Azharul Haq (Eds.), *Water rights, conflict and policy* (pp. 195–220). International Irrigation Management Institute.

Flyvbjerg, B. (2006). Five misunderstandings about case-study research. *Qualitative Inquiry, 12*(2), 219–245.

Gautam, U. (2012). Nepal: Food security, a localized institutional irrigation perspective on public irrigation systems. *Hydro Nepal* Special issue, 95–99.

Government of Nepal/National Trust for Nature Conservation (NTNC). (2009). *Bagmati action plan (2009–2014)*. Kathmandu.

His Majesty's Government of Nepal (HMGN). (1969). *Kathmandu Valley plan*. The Physical Development Plan for The Kathmandu Valley Department of Housing and Physical Planning. Ministry of Public Works Transport and Communication.

His Majesty's Government of Nepal (HMGN) and USAID. (1986). *Kathmandu Valley urban land use policy*. Kathmandu Valley Town Planning Office, His Majesty's Government of Nepal and U.S. Agency for International Development.

International Center for Integrated Mountain Development (ICIMOD). (2007). *Kathmandu Valley environment outlook*. International Center for Integrated Mountain Development, Ministry of Environment, Science and Technology (MoEST) and United Nations Environment Programme (UNEP), Kathmandu, Nepal.

Joshi, N. M. (2018). History of irrigation development in Nepal. In Jalsrot Vikas Sanstha GWP Nepal (Ed.), *Water Nepal: A historical perspective*. Jalsrot Vikas Sanstha.

Kathmandu Valley Development Authority (KVDA). (2015). *Kathmandu Valley 2035 and beyond: 20 years strategic development master plan (2015–2035) for Kathmandu Valley (draft)*. Ministry of Urban Development, Nepal.

KUKL (Kathmandu Upatyaka Khanepani Limited). 2017. *Annual report ninth anniversary*. Kathmandu, Nepal.

Lankford, B., Bakker, K., Zeitoun, M., & Conway, D. (Eds.). (2013). *Water security: principles, perspectives and practices*. Earthscan/Routledge.

Leaf, M. (2011). Periurban Asia: A commentary on "becoming urban". *Pacific Affairs, 84*(3), 525–534.

Ministry of Agriculture (MoA). (2012). *Statistical information on Nepalese agriculture 2011/2012*. Ministry of Agriculture Development.

Ministry of Urban Development (MoUD). (2017). *National urban development strategy (NUDS)*. Government of Nepal.

Muzzini, E., & Aparicio, G. (2013). *Urban growth and spatial transition in Nepal. An initial assessment*. The World Bank. https://doi.org/10.1596/978-0-8213-9659-9

Narain, V., & Prakash, A. (Eds.). (2016). *Water security in periurban South Asia. Adapting to climate change and urbanization*. Oxford University Press.

Nepal Agriculture Research Council (NARC). (2006). *Annual report on Kathmandu Valley vegetable farming*. Vegetable Development Division, Nepal Agriculture Research Council (NARC), Khumaltar, Kathmandu.

Pandey, V. P., Shrestha, S., & Kazama, F. (2012). Groundwater in the Kathmandu Valley: development dynamics, consequences and prospects for sustainable management. *European Water, 37*, 3–14.

Pradhan, R. (2000). Land and water rights in Nepal. In R. Pradhan, F. von Benda-Beckmann, & K. von Benda-Beckmann (Eds.), *Water, land and law. Changing rights to land and water in Nepal* (pp. 39–67). Proceedings of a workshop held in Kathmandu). Freedeal/Wageningen Agricultural University/Erasmus Universiteit Rotterdam.

Pradhan, P. (2012). Results of unplanned programs: drinking water and sanitation systems in Bhaktapur, Nepal. In A. Prakash et al. (Eds.), *Interlacing water and human health* (pp. 381–402). Sage Publications.

Pradhan, P., & Belbase, M. (2018). Institutional reforms in the irrigation sector for sustainable agriculture water management including water users associations in Nepal. *Hydro Nepal, 23*, 58–70.

Rai, M. K., Paudel, B., Zhang, Y., Khanal, N. R., Nepal, P., & Koirala, H. L. (2019). Vegetable farming and farmers' livelihood: Insights from Kathmandu Valley, Nepal. *Sustainability, 11*, 889.

Rana, S., Rijanta, R., & Rachmawati, R. (2015). Multifunctional peri-urban agriculture and local food access in the Kathmandu Valley, Nepal. *Journal of Natural Resources and Development, 5*, 88–96.

Ribot, J. C., & Peluso, N. L. (2003). A theory of access. *Rural Sociology, 68*(2), 153–181.

Roth, D., Boelens, R., & Zwarteveen, M. (Eds.). (2005). *Liquid relations. Contested water rights and legal complexity*. Rutgers University Press.

Roth, D., Boelens, R., & Zwarteveen, M. (2015). Property, legal pluralism, and water rights: the critical analysis of water governance and the politics of recognizing "local" rights. *Journal of Legal Pluralism and Unofficial Law, 47*(3), 456–475.

Roth, D., & Vincent, L. (Eds.). (2013). *Controlling the water. Matching technology and institutions in irrigation management in India and Nepal*. Oxford University Press.

Sada, R., Shukla, A., & Shrestha, & Melsen, L. (2016). Growing urbanization, changing climate, and adaptation in peri-urban Kathmandu. In V. Narain & A. Prakash (Eds.), *Water security in peri-urban South Asia. Adapting to climate change and urbanization* (pp. 147–186). Oxford University Press.

Shrestha, B. (2011). *The land development boom in Kathmandu Valley*. International Land Coalition, CIRAD, College of Development Studies. [Online] http://www.landcoalition.org/sites/default/files/documents/resources/CDS_Nepal_web_11.03.11.pdf

Shrestha, P. (2007). *Anantalingeswor: A cultural study*. Lusa Press.

Shrestha, N., Lamsal, A., Regmi, R. K., & Mishra, B. K. (2015). *Current status of water environment in Kathmandu Valley, Nepal* (Water and urban initiative working paper series number 03). United Nations University. [online] URL: http://collections.unu.edu/eserv/UNU:2852/WUI_WP3.pdf

Shrestha, A., Roth, D., & Joshi, D. (2018). Flows of change: Dynamic water rights and water access in peri-urban Kathmandu. *Ecology and Society, 23*(2), 42.

Shrestha, S., & Shrestha, B. K. (2009). The influence of water in shaping culture and modernisation of Kathmandu Valley. In J. Feyen, K. Shannon, & M. Neville (Eds.), *Water and urban development paradigms* (pp. 105–114). CRC Press.

Simon, D. (2008). Urban environments: Issues on the peri-urban fringe. *Annual Review of Environment and Resources, 33*, 167–185.

Tiwari, S. R. (1999). *Kathmandu Valley urban capital region and historical urbanism historical environment management: Lessons from history*. Paper prepared for presentation to 13th biennial conference of Association of Development Research and Training Institutes of Asia and the Pacific (ADIPA), on the theme of 'Managing Asia-Pacific Mega Cities: Policies to Promote Sustainable Urban Development in the 21st Century', 29 November–1 December 1999, Bangkok, Thailand.

United Nations Human Settlements Programme (UN-Habitat). (2015). *Cities and climate change initiative: Kathmandu Valley Nepal – Climate Change Vulnerability Assessment*. Kathmandu Valley, Nepal – Climate Change Vulnerability Assessment. United Nations Human Settlements Programme.

Upreti, B. R. (2000). Community level water use negotiation: implications for water resource management. In R. F. Pradhan & K. von Benda-Beckmann (Eds.), *Water, land and law. Changing rights to land and water in Nepal. Proceedings of a workshop held in Kathmandu* (pp. 249–269). Freedeal/Wageningen Agricultural University/Erasmus Universiteit Rotterdam.

Water and Energy Commission Secretariat (WECS). (2011). *Water resources of Nepal in the context of climate change.* Government of Nepal. URL: http://assets.panda.org/downloads/water_resources_of_nepal_final_press_design.pdf

Yin, R. K. (2009). *Case study research: Design and methods.* Sage Publications.

Zeitoun, M., Lankford, B., Bakker, K., & Conway, D. (2013). Introduction: A battle of ideas for water security. In B. Lankford, K. Bakker, M. Zeitoun, & D. Conway (Eds.), *Water security: Principles, perspectives and practices* (pp. 3–10). Earthscan-Routledge.

Zeitoun, M., Lankford, B., Krueger, T., Forsyth, T., Carter, R., Hoekstra, A. Y., Taylor, R., Varis, O., Cleaver, F., Boelens, R., Swatuk, L., Tickner, D., Scott, C. A., Mirumachi, N., & Matthews, N. (2016). Reductionist and integrative research approaches to complex water security policy challenges. *Global Environmental Change, 39*, 143–154.

Chapter 4
Public Lives, Private Water: Female Ready-Made Garment Factory Workers in Peri-Urban Bangladesh

Deepa Joshi, Sadika Haque, Kamrun Nahar, Shahinur Tania, Jasber Singh, and Tina Wallace

4.1 Introduction

Bangladesh is South Asia's fastest emerging economy with a Gross Domestic Product growth rate of 7.11% (World Bank, 2017a). This is a significant achievement for a predominantly low income country. The Ready Made Garment (RMG) industry, the country's largest export sector, is a key contributor to this economic growth trajectory and is also the country's largest employer of women. Of the over 4.2 million people working in the RMG industry, around 80% workers are women (CPD, 2018). It would not be incorrect to say that female RMG workers are key drivers of the Bangladeshi economy (see Rahman & Siddiqui, 2015).

How has this employment changed the everyday lives and livelihoods of this essentially female work force? While researchers acknowledge the formidable challenges for Bangladeshi women working in the RMG industry (Feldman, 2009; Kabeer, 1999, 2000, 2001; Kabeer & Mahmud, 2004a, b), there are nonetheless claims that the mass employment of women in the RMG industry has led to significant gender gains (World Economic Forum, 2018, 2019). It is argued that entrenched

D. Joshi (✉)
International Water Management Institute (IWMI), Colombo, Sri Lanka
e-mail: deepa.joshi@cgiar.org

S. Haque · S. Tania
Bangladesh Agricultural University, Mymensingh, Bangladesh

K. Nahar
University of Manitoba, Manitoba, Canada

J. Singh
Coventry University, Coventry, UK
e-mail: ac5866@coventry.ac.uk

T. Wallace
Oxford University, Oxford, UK

V. Narain, D. Roth (eds.), *Water Security, Conflict and Cooperation in Peri-Urban South Asia*, https://doi.org/10.1007/978-3-030-79035-6_4

gender barriers, in particular cultural restrictions on women's physical and social mobility and ideological perceptions that women cannot or should not earn income have been overcome (Kabeer, 2001). RMG women workers in Bangladesh are also said to "have a (relative) autonomy [...] a greater bargaining power [...] expanded choices and gains in power" (Hossain, 2012, p. 3) and "an enhanced ability to formulate choice[s] and act upon those choices" (Souplet-Wilson, 2014, p.1). It is no surprise then that the RMG industry in Bangladesh is widely claimed to have empowered Bangladeshi women (World Bank, 2008, 2017b). The World Economic Forum identifies Bangladesh as the most gender equal country in South Asia and as having surpassed developed economies as well in furthering gender equality (see World Economic Forum, 2018, 2020).

However, the RMG industry in Bangladesh is also infamous globally for two major events: a fire breakout in the Tazreen factory in Ashulia District in peri-urban Dhaka in 2012, where 117 workers were confirmed dead and over 200 were injured; and the collapse of the Rana Plaza building complex in 2013 in another peri-urban suburb, Savar, with a death toll of 1134 and 2500 injured. Much has been written about these incidents and also about subsequent efforts to promote and ensure "safety-at-work" initiatives.

The political economy of the RMG industry in Bangladesh makes for a perfect case to understand how globalization intersects with gender. Of course, it depends on how these links are analysed or rather, who is analysing these intersects. The World Bank's World Development Report (2012, p.xxi) notes that "forces such as trade openness [...] spread of cheaper communication and technologies [...] connect women to markets and economic opportunities [...] reshaping attitudes and norms about gender relations, and encouraging countries to promote gender equality." This analysis explains how Bangladesh ranks at 50 among 149 countries in the World Economic Forum's 2020 Global Gender Gap index report, where a key indicator is economic empowerment.

Feminist researchers critique these measures, as well as the narrative that asserts that any wage employment for women translates into their empowerment. Ruth Pearson (2007, p.117) explains that trade liberalization has reiterated rather than reversed gender inequalities: by making poor women the cheap workers of globalized value chains, even as these women combine the burdens of poorly paid productive work with unpaid domestic work responsibilities, ensuring that "being exploited by capital" is virtually the fate of poor women in today's global economy.

In 2018, Shafiqur Rehman wrote of the imbalances in relation to gender equality and the RMG economy in Bangladesh,[1] highlighting two key issues. First, Rehman argued that the work opportunities for poor women in the RMG industry are poorly paid and "labour-intensive", with a "lack of learning and upgrading of jobs" where workers have "short shelf-lives" and are "unceremoniously terminated (with no

---

[1] https://www.dhakatribune.com/opinion/op-ed/2018/12/24/imbalance-in-economy-and-politics-of-bangladesh

post-work benefits) after the end of the productive years" at work. In other words, this is not "decent work" and does not offer "living wages" and meet other decent work conditions (see International Labour Organization et al., 2016). Secondly, because "more than 85% of the total exports" in Bangladesh "is just from the RMG industry" and because the export industry runs the country's "'miraculous' economic growth [….] the industry creates a narrow business elite that can easily collude with the government for preferential treatment". In a situation where the nation's GDP is buoyed by the RMG industry, critiques of the economic and/or environmental externalities of the industry have made little dent on what happens on the ground.

In this chapter, we discuss how poor women in Bangladesh not only service the RMG labour market on unfavourable terms and in challenging work contexts, but also function in a social and cultural setting where gendered norms dictate that, even as women work outside the house, they are still fully responsible for unpaid care and domestic work. This happens in a context where, while the women pay almost a third of their wages for small rooms in crowded tenements, sharing poorly maintained and overtly crowded toilets and bathrooms, they are unable to question the landlords on the conditions of their rents and living conditions, especially also water availability, use and access.

We understand women's empowerment as their ability and agency for "transforming power relations" at scale, so as to "define change for themselves and negotiate change" individually and collectively, at scale and on their terms (Sharma et al., 2007, p.11). This conceptual understanding of empowerment defines that "the material and non-material dimensions of all our lives are inextricably intertwined" and therefore it is not possible to segregate the economic, social and political dimensions of empowerment (ibid, p.12). Our findings mirror what has been reported elsewhere—an overwhelming "feminization of responsibility [of both domestic and productive work] and/or obligation" resulting in "a growing unevenness in male and female inputs to household economies" in situations where "women and girls have limited power to (re)negotiate this intensified reliance" on their time, labour and lives (Chant, 2016, p. 3–4). Building on what we observe through a gender-water-work focus, we critique what is positioned and claimed as economic development and women's (economic) empowerment.

In the sections below, we first explain our approach of a critical participatory ethnography. This methodology section is followed by a conceptual overview of the political economy of the RMG sector and how this impacts water inequalities. We then discuss in detail the research findings—where we analyse the meanings and experiences of privatized waters and the public lives of women RMG workers. This section is then followed by a final discussion and conclusion.

4.2   Methodology: Our Framing of Their Stories. Ethnography in an Unequal World

Informed by a feminist political economy framework and guided by critical participatory research methods, our aim was to facilitate some 30 female RMG workers
working and living in Bhadam to reflect on their everyday lives in the public and
private spaces where they work and live. Bhadam (Fig. 4.1), the location for this
research, a small settlement (locally known as *mouza*) in Gazipur District in Dhaka
Division, is one of many such peri-urban settlements, where RMG workers live by
the hundreds along the Dhaka-Gazipur highway. Gazipur, which used to be part
wetland, part paddy fields on the banks of the Turag River, is now the heavily indus-
trialized heartland of the RMG sector.

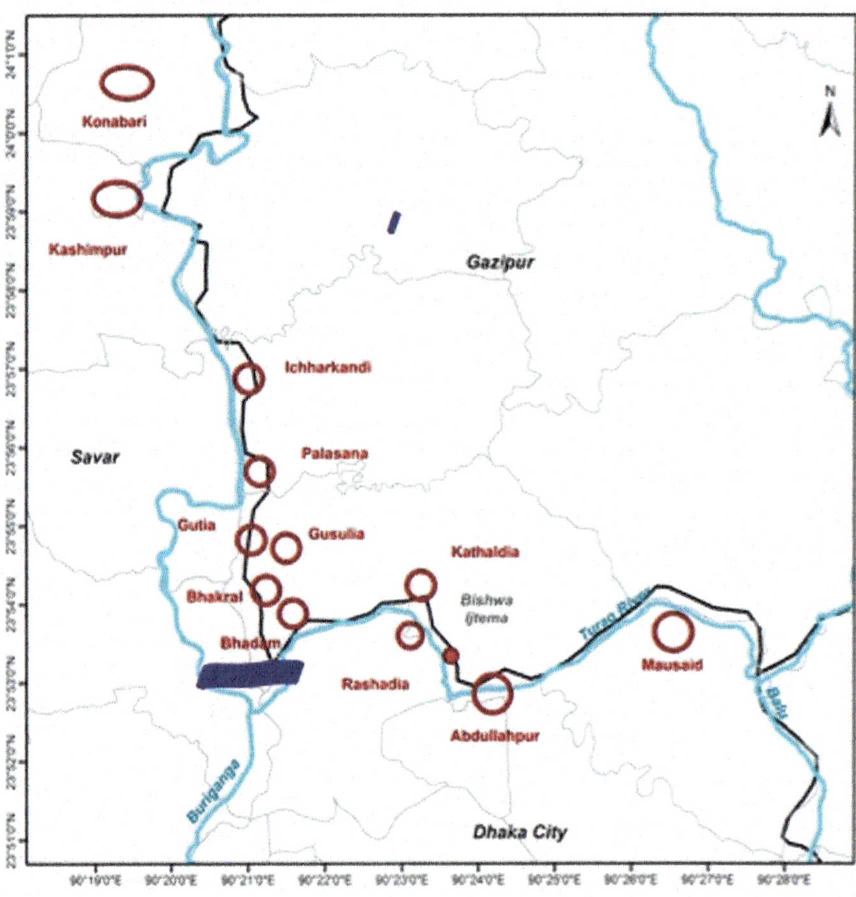

Fig. 4.1 The study area

Our focus was to enable women working in the RMG factories to voice their experiences, hopes and challenges in a context not driven by an outsider research agenda to extract data. Thus, while this research project was essentially about water, we did not emphasise this in our engagements and discussions with women RMG workers. If water was a key concern to the women we spoke to and interacted with, it would, we anticipated, emerge as a key topic of discussion. And indeed, water did emerge as a key challenge, with other issues of gender-power disparities and masculinities, both inside and outside the home, in private and public domains.

Despite the fact that, in Bhadam, the factories and the one-room accommodations the workers rent are not too far away, getting an insight into people's lives by "deeply hanging out" (Geertz, 1998) was immensely challenging. Most of our interactions with the female RMG workers happened in the small, cramped rooms where they lived with their families, on Fridays, the day off for RMG factory work. However, as we noted, factory work also occasionally happens round the clock to meet specific seasonal demands of customers in the North. The key challenges were the unavailability of time to meet and speak, and the lack of space for discussions. The living quarters—rooms divided by tin shutters—were in some cases just behind the factory walls. The hum of generators and other factory machinery made it difficult to speak to or hear each other—the public work place seamlessly flowing into the domestic space, which is anything but private.

Two of our researchers (and co-authors) lived in Bhadam in such rented rooms for over 8 months and experienced first-hand some of the challenges faced by RMG women workers. They were evicted from their rental accommodation three times, because the male landlords decided in each case to oust them arbitrarily, without explanation or notice to leave. We thus came to understand how RMG workers and their families experience the power and authority of male landlords. While the rent workers pay is almost a third of their wages for such living arrangements, they are completely unable to challenge forced eviction or rent, water and electricity charges, water rationing, poorly maintained toilets and bathrooms conditions, and the overall exercise of control of space (Fig. 4.2).

In this challenging environment, we met and spoke to over 30 female RMG workers as well as several male members of their family—individually and sometimes in a group, over a period of a year. Twelve of these women reflected on their lives and their work with a local street theatre group called *Bot Tala* (under the Banyan tree). These reflections were later translated into a theatre play, which has been co-produced with these women into a video.[2] Overall, the stories of the 30 women were stories of deep poverty, lack of choice, distress migration, challenging living and working conditions, hard tedious work at home and at the factory, tiredness, low status in the community and, of course, an endless battle to secure enough water for all household purposes.

It took significant time to win the trust of these 12 women—many were shy, lonely and also wary of sharing their life experiences with others. The gradual

---

[2] For the video, see https://youtu.be/7GxkzNWRVlM

Fig. 4.2 Housing of 20 families of female garment industry workers. (Photo Sadika Haque)

nature of interactions and the space to discuss and share everyday lived experiences at their pace were key in engaging with them. In trying to be mindful of the tensions and contradictions of power and privilege in doing research, our project moved slowly. Our co-authors often took on the role of surrogate care givers and educators of the children of the women we wanted to speak with. During the short-term, mission-driven purpose of our engagement we were glad to be useful in some ways. For many literate RMG workers, the fact that they were raising their children into illiteracy—there are no public schools in Bhadam—was of deep concern. We tried, not always successfully, to not make compulsions or demands on their engagement. There was a lot shared over time, much of what they cared about deeply is not possible to fully detail in this chapter: their childhood memories; experiences of migration; compelling poverty in the rural areas; the pain of separation from children left behind in the villages or all day while at work; aggression and abuse by men at home and men at work; physical and emotional challenges in their marital relations due to, among other issues, time stress; their isolation from wider community support; problems with not having access to schools and healthcare; having no recreational spaces or time to relax—and much more. The focus in this chapter is on their experiences as urban workers in the RMG industry and as working wives living in urban slums where housing is costly and overcrowded, where water supply is inadequate and unilaterally determined by male landlords, where the gross pollution of surface and groundwater and the overall environment go unregulated, and where the lack of time impacts everything that is expected socially of a good, virtuous and desirable wife and mother.

As researchers working in Bangladesh for a long time, we have often grappled with the politics of "researching" poverty and gender (Joshi et al., 2011). Poor women in Bangladesh may lack many things, but research attention is something that they are often disproportionately subject to. Many researchers, including us, claim to be "doing" participatory research, guided by "a feminist ethnographic methodology, which gives careful attention to diverse experiences and voices by heeding the constraints, opportunities and challenges in women's lives, and looks for thick description, specific data and stories that are profoundly gendered and relevant to the context" (Sultana et al., 2013, p.5). And yet, working on issues of gender, water and poverty in Bangladesh, we have been acutely aware of the tensions and contradictions between a scientific obsession with data—also called the "mission-creep of scientific knowledge processes" (see Wakeford & Sanchez-Rodriguez, 2018)—and the purpose, value and relevance of such data for those who are "researched". While participatory, ethnographic methods may deliver nuanced, granular insights into complex ground realities, to assume that such research can or will shift the praxis of social injustices (Palmer, 2001) is deliberately naïve. We argue here that "deeply hanging out" with the marginalized does not easily qualify as being inherently political or participatory. Escobar (2007, 2015, 2016) argues, that, regardless of the focus on the sub-altern, if the goal is to contribute to established Eurocentric knowledge systems—then, no matter how critical the methods and approaches, they do not necessarily qualify as being epistemologically transformative. Transformative, transdisciplinary research requires "changing not just the content but the very terms of the conversation" (Escobar, 2007, p. 191).

It is, therefore, important to acknowledge that, while project funding enabled us, a group of female and male researchers from Northern and Southern institutions, to try and *deeply hang out* with female ready-made garment (RMG) factory workers in peri-urban Bangladesh, nonetheless, despite our best intentions of being participatory, the findings we discuss here can only qualify as "hybrid research—occasionally but not always facilitating 'participants' to make sense of [and challenge] their real situations and contexts" (Wakeford & Rodriquez, 2018, p.26).

4.3 Public Water Privatized

Before discussing in further detail the intersection of paid work and gendered domestic water responsibilities of RMG women workers, it is important to understand how the political economy of the RMG industry in Bangladesh has shaped the political economy of water. As we discussed above, the RMG industry in Bangladesh is acclaimed for accelerating productivity and the nation's GDP as well as the empowerment of the industry's women workers. And just as the everyday lived experiences of the women workers at work and at home are silenced in outsider expert claims of (their) empowerment, the environmental degradation as an outcome of industrialization and growth is also grossly overlooked.

Hossain and Samad Khan (2017) stress that the industry's ecological footprint is alarming. While the industry's entire value chain is "highly water intensive", what is a more immediate concern is the unregulated mining of groundwater by the industry (at no costs) and the release of untreated effluents containing several hazardous dyes and heavy metals. The RMG sector, including textile dyeing, is classified officially as a highly polluting "Red Category" Industry under the 1995 Environmental Conservation Act (ibid; 438). The industry is responsible for an average of "86.15% of total grey water footprint for the country" (ibid; 438, 447). Ironically, as of May 2019 Bangladesh also reported the highest number of Leadership in Energy and Environmental Design (LEED)-certified garment factories globally: 24 according to the United States Green Building Council (USGBC).[3] However, this number, compared to the absolute numbers of large and small RMG factories identified in total as 4621 in 2018–19 by the Bangladesh Garment Manufacturers and Exporters Association,[4] explains why, as Sakamoto et al. (2019, p.1) note, the industry's supposedly "great success comes with great environmental deterioration".

Ignoring the overall "agricultural water consumption for cotton farming", various textile industries "in and around Dhaka may consume as much groundwater as goes to all of Dhaka's residents" (Hossain & Samad Khan 2017, p.437). More importantly, the water pollution has far-reaching impacts: "vegetable and fruit samples collected from around Savar, Dhamrai, and Tongi (other peri-urban locations where the RMG workers live) show the presence of textile dyes, industrial wastes and effluents containing heavy metals [...] all regularly released into the already heavily polluted river water, which is still being used for irrigating paddy and vegetable (spinach, tomato and cauliflower) fields in industrial areas in Gazipur and Keraniganj" (ibid, p.6). The RMG workers we met in Bhadam knew that the Turag River is too polluted to even wash clothes, but they can do little to protect themselves from the pollution in the water they drink and the food they eat.

The study region, including Bhadam, is not served by any water supply provided by the state. Land- and home-owning residents of Bhadam, who are landlords to large numbers of migrant RMG workers, make personal investments to draw groundwater for domestic use from tubewells by means of submersible pumps. There are no fees (to the landlords) for using groundwater, but there are significant costs for buying, setting up and maintaining the water pumps, and the electricity to run the pumps. This explains why water availability is a contentious issue between the landlords (usually men) and tenants. At all times and regardless of the rental arrangements, landlords decide the timing and frequency of running the submersible pumps. Women RMG workers are mostly getting back from work at a time of the evening (7–8 p.m.) when the water storage tanks have run dry. While the women need water for drinking and cooking, there is no option but to wait till the next refill or to request the landlord or his family members to extend a favour. These requests are often not taken well and create an enormous anxiety each day. None of the RMG workers in Bhadam own land or have their own homes. The lack of political capital as migrants, the lack of land, homes, financial capital or other assets of their own

---

[3] https://www.thedailystar.net/business/news/bangladesh-has-highest-number-green-garment-factories-1749016

[4] http://www.bgmea.com.bd/home/pages/TradeInformation

means that the migrant RMG workers cannot sink water pumps. It is worrying that migrant workers lack these resources, regardless of the length of stay in these areas and work in the factories. In sum, regardless of how long they have migrated, they are perpetually migrants, lacking political, social and financial capital.

In a rare, one-off incident in a nearby settlement, a group of migrant workers living in what was essentially a slum-like situation in Gazipur, had tried petitioning local authorities and had been allotted a submersible pump from the local office of the City Corporation. Several people in the community lobbied with higher-ranking government officials in the same institution and managed to take back the pump. Some form of politico-legal identity might have restored some voice but, as we learnt, most RMG workers also do not aspire for a local politico-administrative status in the peri-urban fringes where they live. Not a single respondent we met had officially registered as local resident (i.e. an eligible voter). When we asked why, every single person, male and female migrant RMG worker, said that this was because they want to go back home one day. It is also true that most workers, especially women, have neither the resources nor the time to pursue these changes.

However, even if this politico-legal barrier was negotiated, it is not possible for the migrant workers to put together the financial resources to access water. Often it takes three to four less affluent landlords to pool resources to invest in one submersible pump. As the wife of a landlord explained:

> We, three house owners, are the joint owners of one pump. When the pump became dysfunctional for a week, we were all, including the tenant workers, collecting water from the nearby mosque. As my husband could not manage money to repair it, I have sold my ornaments and given him Tk. 45,000 (USD 525). Each of the owners is contributing the same amount to set up a new water pump.

Focusing on water, Sultana et al. (2013) have written that, since the 1980s, neoliberal economics has framed the country's politics, in the process also determining the institutional mandate of the water sector, as well as notions of citizenship and rights to basic services. While policy defines water as a public good, in practice water is a private commodity in Dhaka city and the surrounding peri-urban areas. The rate of growth and expansion of the city undoubtedly constrains the city's (Dhaka) Water Supply and Sewerage Authority (DWASA), which as it appears is also fighting external agendas of corporatization (i.e. privatization) introduced as "water reforms" by agencies like the Asian Development Bank (ADB). In Dhaka, DWASA has, however, resisted privatization, even though the ADB continues to emphasize "a time-bound action plan for private sector participation" (Asian Development Bank, 2017).

Interestingly, the ADB report, titled, "The Dhaka Water Services Turnaround" notes that DWASA executives and engineers were initially resistant to delivering water to urban poor slum dwellers, arguing that this would "affect the pressure in other areas of the system and hurt paying customers" and would "turn *those water thieves* (the urban poor in slums) into regular customers" (ibid; p.32). A change in attitude was apparently possible because the ADB was able to engage local NGOs to take fiscal responsibilities and became intermediaries between "the utility and the communities" (ibid; p.34). We spell out these issues here because, unlike in urban slums, there are no legally informal areas, i.e. slums, here. These are well organised migrant-tenant settlements. We also found no local NGOs looking into issues of exclusion and marginality in Bhadam or in the neighbouring areas. As we understand, organisations like UNICEF are engaged in improving WASH services in factories; and NGOs like Action Aid work with RMG workers, including women, to support and extend worker rights. However, both were not known or recognised by the people we met and spoke to in Bhadam—the women RMG workers, their families, local residents (i.e. landlords) or those who had managerial positions in the factories. This is hardly surprising: Bhadam lies some six kilometers away from the main road, and one cuts through several such settlements before one arrives here.

It is important to add here that the Constitution of Bangladesh abides by and acknowledges the obligation of the nation state to ensure that basic necessities of life such as food, clothing, shelter, education and medical care are accessible to all citizens. Water, however, is not listed as one of the basic necessities of life. This oversight is perhaps explained by the fact that the 1999 National Policy for Safe Water Supply and Sanitation, declares water as owned and controlled by the state and to be managed by public, private and/or community institutions, as deemed appropriate by the state (Ministry of Water Resources, 1999). Similarly, the Government of Bangladesh's (GoB) 2005 Poverty Reduction Strategic Papers (PRSP) and Bangladesh's Water Act 2010 speak of the citizen's right to water, but not the state's obligation to supply safe water. It is for all these reasons, that DWASA is commended for its efforts to resist privatization and persist as a public water institution (Transnational Institute, , 2004). However, as we discussed above, public water institutions are not necessarily inclusive.

More recent shifts in Bangladesh' water and sanitation policy aim to target the hard to reach areas and people, i.e. the poorest and most disadvantaged. This turn-around in policy is spelt out in the 2010 Sector Development Program document which, in contrast to earlier policies and guidelines, states the obligation of the national government to provide safe drinking water to all its citizens. The SDP document also outlines the need for equitable subsidies and for local contextual issues to inform water development decisions, including the consideration of social values, cultural practices and technical appropriateness of the design and development of water services and infrastructure. In principle, these policies could inform the ambitiously planned transition of the DWASA to WASA, a 23-year master plan—the 2035 Greater Dhaka Water Supply and Sanitation Plan. Among other issues, this plan aims to ensure water supply and sewerage to the currently unserved peri-urban regions of Savar, Tongi, Keraniganj, Purbachal and Gazipur. It is unclear how these plans will take into account the challenging problems of skewed landlord-tenant relations, which we describe below. Further, what matters is not just provision of water infrastructure and services but, as Sultana et al. (2013, p.2) have argued, also recognizing that "place-based politics determine […] rights and injustices" in the socio-political "fabric". In other words, there cannot be equitable water provision without full rights of citizenship (ibid). And as we discuss below, gender plays prominently in the experience of place, space, identity, citizenship, water infrastructure and governance.

4.4   Living in Bhadam: Women in Masculine Spaces and Places

In the late 1990s, the RMG industry moved from the urban fringes of Dhaka into surrounding rural landscapes, which are today the peri-urban industrial zones including areas like Gazipur, Savar and Narayanganj. The rapid development of the industry coincided with a rampant construction of rental accommodations for migrant workers from impoverished rural areas with little opportunities for income. Depending on the season, monsoon or drier months, it takes about an hour or less to navigate the seven kilometres from the main Gazipur-Dhaka highway to Bhadam. The tarred roads of the highway rapidly give way to small roads, which are essentially soil embankments on either side of wetlands and agricultural lands. It is extremely difficult to drive and even more precarious to walk these roads, as well as the streets in Bhadam in the monsoon, when wet clay soils become slippery pathways.

Bhadam as a place is unnervingly masculine: tall cemented factories fenced by tall barbed wires, manned (pun intended) by male guards round the clock. Heavy duty trucks and warehouse transport vehicles traverse through and often block the narrow temporary roads connecting the factories to the main road. Nearby are clusters of settlements where the migrant workers and Bhadam's more settled residents live—interspersed with a string of small provision stores and tea stalls, where men hang out in their spare time. The only institutional infrastructure other than the factories are mosques—and a significant number of minarets which dot the neighbourhood call out prayers every Friday, and five times a day. The Friday sermons, broadcast through loudspeakers, are particularly interesting. Often when we were there, there was a warning to all, but especially to women, of the impacts of "modernity". Women were reminded to be virtuous, chaste and abiding to their husbands.

4.4.1   At Work

Although RMG work is presented as offering economic as well as social opportunities for women, the experiences of the women we spoke to were different. Distress migration is what pushes women from rural villages into the RMG factories. There were stories of crushing poverty, crises of debt, ill health or death in the family, or sometimes womens' decision to elope with lovers, against the wishes of their families. RMG work is relatively easy to find for women but it provides no contracts, no terms and conditions of work, no scope for growth and advancement, and work tenure is not permanent. RMG workers are obviously poor and perceived as desperate. Lacking other means of a livelihood they are looked down upon by the local (non-migrant) community in Bhadam. Local women do not work in the RMG factories. A few resident local men work in the factories, but always in better paid, less laborious jobs. Even though most of the migrant women working in the RMG factories say that factory work, even with all its limitations, is *still* better than other informal jobs in the area, their everyday experiences are rarely positive. Of the 30 women we spoke with, only two spoke of the workplace as being welcoming, as having presented opportunities for a better life and friendships with other women. For most women, the balancing of poorly paid, laborious work in the factories and domestic housework is enormously challenging.

On a daily basis, if women reach the factory gates late even by a few minutes, male guards can disallow their entry, resulting in loss of work and the equivalent wage for that day. Women have to negotiate for leave, if they or other family members are sick; and often have to struggle to ask for toilet breaks during working hours. The insecurity of their work, their inescapable poverty makes them perpetually scared of being docked without pay for being absent without proper permission. Their relationships with the male guards and male supervisors, who are key liaison persons with the administration, is skewed. There are many stories of manipulative behavior: the more beautiful and young(er) one is, the easier it is to get a job, and a less tiring one. Women and girls considered not so beautiful, or too old, risk not being hired at all—and multiple favours are demanded by the men in power. The sheer numerical strength of women workers has unfortunately not enabled them to challenge the masculine hierarchies at work. Work conditions are changing but not significantly improving for the women: there are now improved WASH facilities inside many factories, but women have little time to use them. Wages have increased but net income or savings have not. There are trade and labour unions and regulations, but these are almost entirely comprised of men. It is still difficult to negotiate leave, there is no maternity leave, working hours are still long, work is often arduous and indiscriminate terminations continue—and the list goes on. Even though women do earn an income, this has not enabled them to challenge the conditions of their work.

During the work day, women get an hour's break for lunch (Fig. 4.3). This is when they queue for the toilets or access drinking water, both of which they avoid during work hours because of the risks of not meeting production targets. Those who live nearby sometimes rush home to check on the children, clean, wash and eat before rushing back to work. The rush and competition continues when they return home in the evening—to cook, clean, wash and be the wives and mothers they were not during the working day. Their insecurity and fear of breaking rules came through in most of our conversations, as did the relentless pressure of targets, trying to fit everything into the day, and the sheer lack of time. Some do not drink to avoid the need for a toilet break. They often talked of intense tiredness and the repetitive nature of the work amidst dust and dirt. They also spoke of often having to accept compulsory overtime work, for which they are paid additionally in most cases.

In December 2018, the average salary increased from BDT 5300 (USD 63) to around 8000 BDT (USD 93). Although this was a significant rise, this is only half of what labour and trade unions had been demanding as living wages (BDT 16,000 or USD 190). The current salary equals just over three USD for over 8 h of work per day, 6 days a week. Almost a third of the wages is spent on rent for a small one-room house. Often, as soon as the salaries increased, the house rents also increased

Fig. 4.3 Factory workers going for lunch. (Photo Sadika Haque)

proportionately. We know this not only from the women, but also because the rent for the room we were renting ourselves increased proportionate to the increase in wages during this time.

The conditions of work described by the women do not convey a sense of empowerment. This does not imply there are no benefits or gains from paid work. Most women express a sense of pleasure and pride in being able to earn an income, contributing [primarily] to the household upkeep and sending money home for the children left behind. Yet very few women can determine how to spend the income they earn or spend money on themselves; while everyone speaks of trying to save money to return home, very few are able to do so.

4.4.2 At Home

In Bhadam, the type of rental accommodation and the quality and quantity of water for domestic use and WASH facilities are influenced by what workers can pay for a room (varies from BDT 2000–5000). Regardless of what they pay and therefore where they live, there are no contracts and no mutually agreed terms and conditions of rent, while eviction is an everyday threat. The landlord unilaterally determines rental conditions; there are no norms, no rules, no regulation of who pays what and why. Additionally, there are intermittent cuts in the water supply, random extra changes and constant demands of sexual favours and/or threats. The landlord-tenant relationship is a key indicator of disproportionate power exercised by the male land-lord, regardless of the length of stay in the community of RMG workers.

It was reported that owning a fridge, a TV, having more children, a larger family, or occasional guests can result in increases of the rent amount. Often any additional costs of repairs, maintenance, and high electricity bills are passed on to the vulnerable tenants. Notice to leave can be given at any time. When we (co-authors) were evicted, it was immediate; the landlords gave us less than an hour to leave. The RMG workers, especially single women, live with this threat of evictions every day. Space, place and basic infrastructure and services are not issues that tenants can voice opinions about. The demand for rental accommodation is high and there is no incentive for landlords to be reflexive, though a few migrant men, including husbands of RMG workers, protest these injustices.

The overcrowding in the living arrangement is shocking: sometimes over 100 people live in 20 or 25 rooms, equipped with two toilets, two taps, one or two bathrooms and a limited number of shared kitchen spaces and cooking stoves. The toilets and bathrooms are not segregated by gender and the lack of privacy is burdensome and shaming, especially for women. Lack of water due to this intense overcrowding is common. Many women start their day at 3 to 4 am in the morning to wash and use the toilets and kitchens before the long queues build up later in the morning. Many of the stories shared were related to the lack of time to access the facilities, the shortages of water at peak times, how broken taps can be left unfixed for days on end, the lack of privacy and self-respect, and extremely dirty toilets and washrooms that they share with the other tenants. The women also routinely talked about how hard it is to get to work in time, especially during the rains, with slippery roads and the overspill of waste water on the pathways.

There is little space for migrant RMG residents in general, but particularly for the women, to influence decisions around water. Much of the video produced focused on the distress caused by the shortages of drinking water, the refusal of landlords to turn on the pumps to meet daily water needs, river water that is polluted and unusable, and the long queues for the toilet and bathrooms, which is especially problematic for women who need to reach the factories in time. Landlords include water in the rent, but they limit water availability by closely monitoring the use of the water pumps. Attempts by women tenants to approach male landlords to request additional water (switching the pump on) is looked upon by his family as a way of these women to make sexual advances to a desirable, powerful male. During the research we noted that fairly straightforward requests by these women often result in aggressive responses and/or physical violence by the wives and extended families of the landlord. RMG women workers who speak up or argue are seen as opinionated and immoral, and routinely harassed and even beaten by the wives of the landlords or the landlords themselves. A female RMG worker aged 15 explains:

> one Friday, when I was at home, I noticed that there was no water in the tank. At the very moment, the son of the house owner, aged 17 was passing by. I asked him to request his mother to switch the pump on. Immediately afterwards, the wife of the house owner came out and started beating me with her sandal, expressing anger on how I had dared to speak to her son about water. 'Why did you do this? You should not speak with my son. Tell me if you need anything'.

Mimi, who is young, beautiful, deserted by her husband and childless is precisely the sort of woman that the wives of landlords fear and the landlord himself and his male managers see as easily accessible for sexual harassment.

4.4.3 Gendered Household Water Burdens: Not Negotiable

Despite working long days of 8–10 h, 6 days a week, when women return home they resume the role and burdens of being mothers, wives, daughters and/or daughters-in-law, even when they are primary household income earners. Managing domestic tasks is squarely women's responsibility. Only three out of 30 women said their husbands help them in any way. One of this group of three said:

> Sometimes if my husband comes earlier from office than I and is hungry, he will cook some rice, collect some water and sometimes even wait to eat, till I arrive". Another said: "my husband always helps in all the work, we cook together, cut fish, wash clothes together. When we were in the village, my husband also tried to help me but my mother-in-law did not allow him to help. This has changed after coming to Dhaka: here, he can help me as much as he wants.

And yet, she is still, "very afraid of him", of annoying him, of losing this good fortune that so many others around her do not have. The majority of women felt they had no way to challenge their traditional gendered responsibilities in the home. For most women, the situation at home is unbelievably challenging:

> I had to do overtime work at the factory yesterday night and came home at about 2 am. On getting home exhausted, I found our room in a terrible mess. I cannot ever imagine that he will cook food for me. I found that he had not even washed his *lungi* (lower body wrap for men) after taking a bath, even though I knew he was at home since the afternoon. I washed it late at night, because he needs it for work the next day and if it is not clean or dry, he will beat me. Actually I cannot express my feelings at that moment. I just blame my fate for such a husband.

Another woman said:

> I have been suffering for a few days with fever and cold. With this physical condition, after getting up, I do all the work, I have washed the bathroom and taken a bath. Today is Friday. My husband had brought fish and chicken from the market in the morning and gone out. He has not returned yet—it is 12 noon. I have just finished cooking this for lunch, even though, I have not had anything to eat since I woke up. Despite being ill, I have to do all the work, he doesn't help me in household work. I cannot ask for his help even though he is a good husband. He does not beat me.

For another, "getting help from my husband is beyond my imagination. My husband has another wife and sometimes he is coming to me just to get some money".

On Fridays, women may get up a little later, but face domestic work that has accumulated over a week: cleaning their single-roomed homes, cooking lunch, washing and drying clothes, bathing children and cleaning the toilets and bathrooms. While women are barely seen in the streets, men congregate in tea-stalls and barber shops on Fridays and after work hours in the evening to get a shave or head massage, to smoke and sit by the sides of the road, and in local mosques on the call of prayers (*Azaan*).

Most of the 30 women we met and spoke to at length reported handing over their entire wages to their husbands or at the very least giving them some "pocket money". Many men increasingly threaten that, if the women cannot pay more "pocket money", they will be abandoned for women who can and will pay more. Some of these men even tell their wives: "I will come back—when you can pay me more than what I am given now [by another woman]". We heard countless stories of men both young and old, addicted to narcotic substances, appropriating their wives' earnings through aggression and violence. At home, at work and in the community, the women continue to be at the mercy of men; and despite the crowded social environment at work and in the community, they are deeply isolated and alone.

From the women, we also heard of an ominous practice, which we term, the "floating husband" syndrome. The social need for a husband has resulted in some women (whose husbands have deserted them) investing in men, who float around different households, staying a few days, collecting some money, being "looked after" with good food, rest, maybe more. In the eyes of the wider community, the women have a husband, who is, as they say, "working and living elsewhere". Even in such situations the women usually do not "leave" their men. The deep-rooted culture of women's inequality means they still cannot easily live alone or abandon their husbands. The skewed relations result in several women being abandoned by their husbands, having husbands who spend their hard earned income on drugs, gambling or other women.

Our findings reveal that women holding paid work have not really been able to negotiate the sharing of domestic burdens with men. Instead the domestic water responsibilities of washing, cleaning, cooking are readily handed over to elderly mothers and younger (not yet working) daughters. Young girls from age five to six onwards routinely assist with the household work and rearing younger siblings (Fig. 4.4). As many women explain there is not much else to do in Bhadam. There are no schools, public or private; open spaces for play need to be found in the narrow roads where trucks ply day in and out, or in the polluted, toxic backyards of factories and the banks of the Turag River.

Further, gendered identities, roles, responsibilities, privileges and barriers have hardly changed among the RMG (male and female) workers we met in Bhadam. Gendered relationships are increasingly frayed, with men expecting and demanding their wives—and occasionally mothers and sisters—to provide both domestic comforts and material privileges. Just like at work, at home too coercive masculinities are the ways used to control the women. Even though so many women report of physical violence and aggression, very few feel able to stand up to or counter this behavior.

Heath and Mobarak (2014, 2015) noted a lowering of fertility rates—i.e. reduced rates of child birth among paid RMG workers—and considered this to be an indicator of their empowerment. Such measures fail to take into account the fact, that there are no easily accessible or reliable medical or health services in Bhadam. The time-pressed women RMG workers have a harrowing time dealing with their personal

Fig. 4.4 Girl on her way to a water source. (Photo Sadika Haque)

health situations and their children's health. They also seem to spend very little on their own health. It is very likely that women RMG workers make for a significant proportion of the recent surge in numbers of unsafe abortions in Bangladesh. According to the Guttmacher Institute, in 2014, an estimated 384,000 women suffered complications from clandestine abortions, 1/3 of this number did not get any post-abortion care that they needed, and 2/3 were treated later for complications from unsafe abortions (Hossain et al., 2017).[5]

The stories shared by the women showed a limited power of women and girls to (re)negotiate an intensified reliance on their time, labour, bodies and lives, as has been reported elsewhere by Chant (2016). And, despite the claims of their empowerment, women have greatly subdued expectations of the relations with the men they share their lives and incomes with, of better work conditions and, last but not least, appropriate, affordable water and sanitation infrastructure and services.

4.5 Conclusion

In the claims made for women's paid work being an indication of their empowerment, there is mention of increased mobility, autonomy, voice and negotiating abilities among women RMG workers. The stories told by the women we met are different from these narratives. While women are indeed mobile and out of the house in public spaces, this often comes at disproportional costs to their everyday lives and is also deeply damaging at a personal level.

---

[5] https://www.guttmacher.org/infographic/2017/complications-clandestine-abortion-bangladesh

Most of what we saw and heard when living alongside the women RMG workers in Bhadam, spoke of a pervasive masculinity. There are multiple interpretations and definitions of masculinity. Here we refer to masculinity as identities associated with men in traditionally patriarchal cultures and practices, where male privilege, "aggrandisement and power" within and outside the household is the norm (Chant, 1999, 199). On the other hand, we also interpret masculinity as cultures, practices and values where the privilege and power of a few over others is not uncommon in social, economic and political interrelations where "accumulation and patriarchy" go hand in hand (see Mies, 1986).

It is indeed true that millions of Bangladeshi women are employed. But is this a conscious choice exercised by them or is it a conscious compromise as a result of a precarity that the women have little control over? Do the wages and income from paid RMG work empower women or do they further fuel a culture of capitalism and patriarchy? The decisions that women RMG workers make in continuing work in the factories or staying married with their husbands or the relations they establish with other men as a form of social protection or even staying in Bhadam in the rooms they rent, seem to indicate complexities that cannot be easily explained as choice, freedom or empowerment.

Nussbaum's (2000) capabilities approach allows paying attention to multiple dimensions of disempowerment—economic, social and emotional unfreedoms at work and home and the inability or the conscious choice to not act on these tyrannies. When viewed from such a perspective, it is obvious that the Bangladeshi RMG female workers have enormous "human capabilities": to be, to cope and, in this case, also to fuel the Bangladeshi economy. Unfortunately, there has been little reciprocity to these women by the country, the industry, and by men in their households and the community. Deep-rooted exclusionary behaviors and attitudes are prevalent, not only among the husbands of the female RMG workers and the male landlords of Bhadam, but also persisting institutionally at scale. In the face of such pervasive, structural exclusions, it is ironic to say that poor women in poorly paid work in challenging living and work situations and contexts are empowered. Empowering women RMG workers will require structural changes in the political economy of work as well and water. The question we ask here is, whether a policy focus on inclusion and the expansion of WASH services in peri-urban areas, such as the 2035 Plan, might resolve the challenges we discussed above. We do not feel very optimistic that deep-rooted problems of water insecurity and exclusions will be resolved so easily, especially because multiple factors *disable* both male and female RMG workers from navigating and engaging in public spaces where water management and governance decisions are made by a few powerful men in places like Bhadam.

It is also worrying that new water policy directives are informed by the rather outdated Women in Development (WID) narrative. Put into practice, this would require women in local communities to take the lead in voluntarily managing water supply infrastructure and services in addition to their domestic burdens and paid labour in the factories. The assumption that women can and must take leadership roles in collecting water fees for communal systems and convincing everyone (in the community) to pay is made, because, "Women are (supposedly) better at convincing than men…" (ADB, 2017; 36). In an overall context where economic and environmental injustices and inequalities are a given, where there is an institutional failure to (deliver) public water, and where women's inequality and lack of power is entrenched in everyday social and economic interrelations, it is surely naïve to assume paid work and a small income will transform deep gendered inequalities. When the structural conditions and environments—or what Nussbaum (2000) calls "functioning capabilities" are disabling, individual employment of women alone does not ensure empowerment. In fact, Nussbaum (ibid, 220) argues that in such contexts, women became, as also seen in Bhadam, "mere instruments to the ends of others, as reproducers, caregivers […] agents of a family's general prosperity".

It is precisely in such contexts, that Isha Ray's (2016) argument for gender transformative investments is critical. According to her, this means not leaving the strategic and fundamental human need for water to ambiguous or "voluntary" institutional arrangements including "market forces or non-governmental actors" or to the women themselves, who work full-time. The problem of ignoring the economic value of women's domestic unpaid work has been well argued (UN Women, 2016–2017). However, the positioning of poor women as common-sense solutions to development, for example, how women are more fortuitous in collecting water fees—are counterproductive to empowering women. Ray (2016) argues that it is the state's role not only to provide basic human needs, but equally to protect its citizens and their environmental resources. This calls for reasonable investments in water, sanitation and other basic services which, while often considered "mundane investments are in fact the backbone of a decent quality of life, and yet they remain significantly under-invested in, relative to the global need" (ibid; 2). It is precisely because "systemic issues in the world economy"—including what counts for work and how it can become decent, are overlooked—and because of the fact that women's lives and well-being and empowerment cannot be compartmentalised, that meaningful "empowerment looks unthinkably impractical and ideological" (Fukuda-Parr, 2015; 101).

To conclude, doing this research made us feel dismally dispirited by the fact that, while deep, systemic fault lines of complex inequalities by gender through institutionalized power and politics, exclusion and inequality remain intractable, we and others will continue to "do" more research on the lives of poor women in Bangladesh and elsewhere. Unfortunately, as we know well from having "done" research on women and water for a long time, such research achieves little other than meeting rhetorical and politicized science goals of producing knowledge. Despite our entanglement in the politics and rhetoric of research, gender and water, we conclude here that poor women cannot be targeted and made responsible for transforming a deeply unequal world. Disrupting pervasive, entrenched inequalities by gender will require addressing the root causes of inequality and transforming unequal power relations. This includes transformations in policies, interventions, investments, innovations and the drivers of these change processes, that is: addressing the very political economy of development.

Acknowledgements The design for this research emerged out of a scoping study that was supported by the project "Climate Policy, Conflicts and Cooperation in Peri-Urban South Asia: Towards Resilient and Water Secure Communities" in the framework of the programme Conflict and Cooperation in the Management of Climate Change (CCMCC) funded by NWO (Netherlands Organisation for Scientific Research) and DFID (UK Department For International Development). The research findings presented here in this chapter are from the REACH programme funded by UK Aid from the UK Department for International Development (DFID) for the benefit of developing countries (Aries Code 201880). However, the views expressed and information contained in it are not necessarily those of or endorsed by DFID, which can accept no responsibility for such views or information or for any reliance placed on them. We acknowledge with deep gratitude all the people we met and interacted with in Bhadam and most of all, the women who work in the RMG factories—for giving their time and sharing their experiences without which this paper would not be possible to write.

References

Asian Development Bank. (2017). *The Dhaka water services turnaround: How Dhaka is connecting slums, saving water, raising revenues, and becoming one of South Asia's best public water utilities*. Asian Development Bank.

Chant, S. (1999). Women-headed households: Global orthodoxies and grassroots realities. In H. Afshar & S. Barrientos (Eds.), *Women, globalization and fragmentation in the developing world* (pp. 99–130). Macmillan.

Chant, S. (2016). Women, girls, and world poverty: Empowerment, equality or essentialism? *International Development Planning Review, 38*(1), 1–24.

CPD. (2018). *Declining female participation in RMG sector: CPD Study*, 7 May 2018. Available at: https://cpd.org.bd/declining-female-participation-in-rmg-sector-cpd-study/

Escobar, A. (2007). Worlds and knowledges otherwise: The Latin American modernity/coloniality research program. *Journal of Cultural Studies, 21*(2–3), 179–210.

Escobar, A. (2015). Degrowth, postdevelopment, and transitions: A preliminary conversation. *Sustainability Science, 10*(3), 451–462.

Escobar, A. (2016). Thinking-feeling with the earth: Territorial struggles and the ontological dimension of the epistemologies of the South. *Revista de Antropología Iberoamericana, 11*(1), 11–32.

Feldman, S. (2009). Historicizing garment manufacturing in Bangladesh: Gender, generation, and new regulatory regimes. *Journal of International Women's Studies, 11*(1), 268–288.

Fukuda-Parr, S. (2015). Chapter 4: Food security as if gender equality and sustainability mattered. In M. Leach (Ed.), *Gender equality and sustainability*. Routledge.

Geertz, C. (1998). Deep hanging out. *The New York review of books, 45*(16), 69.

Heath, R. & Mushfiq Mobarak, A. (2014, September, 10). *Manufacturing growth and the lives of Bangladeshi women* (Research briefs in economic policy). Cato Institute.

Heath, R., & Mushfiq Mobarak, A. (2015). Manufacturing growth and the lives of Bangladeshi women. *Journal of Development Economics, 115*, 1–15.

Hossain, L., & Khan, M. S. (2017). *Blue and grey water footprint assessment of textile industries of Bangladesh*. Conference Paper: International Conference on Chemical Engineering. BUET, Dhaka, December 2017.

Hossain, N. (2012). *Women's empowerment revisited: From individual to collective power among the export sector workers of Bangladesh* (IDS working paper, 389). IDC.

Hossain, A., Maddow-Zimet, I., Ingerick, M., Ullah Bhuiyan, H., Vlassoff, M., & Singh, S. (2017). *Access to and quality of menstrual regulation and postabortion care in Bangladesh: Evidence from a survey of health facilities*. Guttmacher Institute.

International Labour Organization, Organization for Economic Cooperation and Development World Bank Group International Monetary Fund. (2016). *Promote decent work*. Shanghai, China: 27–29 April, 2016.

Joshi, D., Fawcett, B., & Mannan, F. (2011). Health, hygiene and appropriate sanitation: Experiences and perceptions of the urban poor. *Environment & Urbanization, 23*(1), 91–111.

Kabeer, N. (1999). *The conditions and consequences of choice: Reflections on the measurement of women's empowerment* (Discussion paper no. 108). United Nations Research Institute for Social Development (UNRISD).

Kabeer, N. (2000). *The power to choose: Bangladeshi women and labor market decisions in London and Dhaka*. Verso.

Kabeer, N. (2001). *Reflection on the measurement of women's empowerment* (SIDA studies no. 3.: Discussing women's empowerment: Theory and practice). SIDA.

Kabeer, N., & Mahmud, S. (2004a). Rags, riches and women workers: Export-oriented garment manufacturing in Bangladesh. In *Chains of fortune: Linking women producers and workers with global markets*. Commonwealth Secretariat.

Kabeer, N., & Mahmud, S. (2004b). Globalization, gender and poverty: Bangladeshi women workers in export and local markets. *Journal of International Development, 16*(1), 93–109.

Mies, M. (1986). *Patriarchy and accumulation on a world scale: Women in the international division of labour*. Zed Books.

Ministry of Water Resources. (1999). *National water policy*. Government of Bangladesh, Ministry of Water Resources.

Nussbaum, M. (2000). *Women and human development: The capabilities approach*. Cambridge University Press.

Palmer, C. (2001). Ethnography: A research method in practice. *International Journal of Tourism Research, 3*(4), 301–312.

Pearson, R. (2007). Reassessing paid work and women's empowerment: Lessons from the global economy. In A. Cornwall, E. Harrison, & A. Whitehead (Eds.), *Feminisms in development* (pp. 201–213). Zubaan.

Rahman, M. H., & Siddiqui, S. A. (2015). Female RMG worker: Economic contribution in Bangladesh. *International Journal of Scientific and Research Publications, 5*(9), 1–9.

Ray, I. (2016). *Investing in gender-equal sustainable development* (United Nations women discussion paper) (p. 23). UN Women.

Sakamoto, M., Ahmed, T., Begum, S., & Huq, H. (2019). Water pollution and the textile industry in Bangladesh: Flawed corporate practices or restrictive opportunities? *Sustainability, 11*(7), 2–14.

Sharma, J., Parthasarathy, K. S., & Dwivedi, A. (2007). *Examining self-help groups empowerment, poverty alleviation, education: A qualitative study*. Nirantar.

Souplet-Wilson, S. (2014). Made in Bangladesh: A critical analysis of the empowerment dynamics related to female workers in the Bangladeshi ready-made garment sector. *Journal of Politics and International Studies, 11*, 359–394.

Sultana, F., Mohanty, C. T., & Miraglia, S. (2013). Gender justice and public water for all: Insights from Dhaka, Bangladesh. *Municipal Services Project (MSP) Occasional Paper, 18*, 1–24.

The Dhaka Tribune: WEF: "Bangladesh most gender-equal country in South Asia". https://www.dhakatribune.com/bangladesh/development/2018/12/19/bangladesh-one-of-the-most-gender-equal-country-in-s-asia (19.12.2018).

Transnational Institute and Corporate Europe Observatory. (2004). *Reclaiming public water! Participatory alternatives to privatization* (Water justice project series 2004/7). Transnational Institute.

UN Women. (2017). *Annual report 2016–2017*. United Nations.

Wakeford, T., & Rodriquez, J. S. (2018). Participatory action research: Towards a more fruitful knowledge. In K. Facer & K. Dunleavy (Eds.), *Connected* (Communities foundation series). University of Bristol/AHRC Connected Communities Programme.

World Bank. (2008). *Whispers to voices: Gender and social transformation in Bangladesh* (Bangladesh development series; No. 22).
World Bank. (2012). *World development report 2012: Gender equality and development.* World Bank.
World Bank. (2017a). *South Asia economic focus, fall 2017: Growth out of the blue.* World Bank.
World Bank. (2017b). *In Bangladesh, Empowering and Employing Women in the Garments Sector*, February 7, 2017. https://www.worldbank.org/en/news/feature/2017/02/07/in-bangladesh-empowering-and-employing-women-in-the-garments-sector
World Economic Forum. (2018, 2019, 2020), *The global gender gap report.* World Economic Forum.

Chapter 5
Digging Deeper: Deep Wells, Bore-Wells and Water Tankers in Peri-Urban Hyderabad

Nathaniel Dylan Lim and Diganta Das

5.1 Introduction

Envisioned as a high-tech city and an engine of growth for the region, Hyderabad has enacted a series of economic reforms in the state's economy to attract the likes of foreign investors. In tandem with the city's growing aspirations, rural land spaces are gradually encroached upon and incorporated into the city's planning boundary. Some of these spaces are then gradually transformed into premium infrastructural spaces which consist of countless high-rise gated urban developments which are coupled with the provision of a robust and uninterrupted supply of basic infrastructural services. While the people who reside in these spaces enjoy the benefits that came with economic reforms and premium infrastructural developments, the local urban and peri-urban poor have been simultaneously bypassed (Das, 2015). The basic needs of the poor, such as the inadequate provision of water and power supply, are neglected. Moreover, peri-urban spaces often fall prey to rapid infrastructure developments encroaching into these spaces, being earmarked for future speculative developments. These encroachments have not only affected the livelihoods of the inhabitants living within the spaces but have also exacerbated existing water-related insecurities.

Massive urban restructuring has placed tremendous stress on Hyderabad's water resources. While Hyderabad is home to both natural and human-made water bodies, such as the Musi River and the Hussain Sagar (Ramachandraiah & Bawa, 2000), the continual expansion of its urban infrastructural developments has led to the encroachment on several water bodies. In particular, the illegal discharge of untreated wastewater has polluted its water bodies and groundwater, rendering them unfit for consumption (Das, 2015; Sen, 2017). Coupled with the growth of massive speculative developments within the city and rising urban population, water demand

N. D. Lim (✉) · D. Das
National Institute of Education, Nanyang Technological University, Singapore, Singapore
e-mail: diganta.das@nie.edu.sg

V. Narain, D. Roth (eds.), *Water Security, Conflict and Cooperation in Peri-Urban South Asia*, https://doi.org/10.1007/978-3-030-79035-6_5

has skyrocketed, placing tremendous pressure on its already scarce supply of freshwater (Vij et al., 2019).

These water-related issues are often cited as a consequence of the city's rapid urban development and expansion. However, it is important to go beyond understanding these issues as a product of urban development. Rather, it is about framing our understandings of water within the larger field of urban political ecology, and acknowledging the complex entanglements between nature and society, as well as the complex power-relations in the study of water (Loftus, 2015; Swyngedouw, 2004). Existing literature has delved into various water-related issues, their causes, and impacts on the local inhabitants in Hyderabad. However, our chapter aims to provide a nuanced understanding of these issues by revealing the perspectives of young residents and migrants in the city through their day-to-day interactions with the water they use. Based on ethnographic fieldwork conducted in Hyderabad's urban and peri-urban areas, this chapter intends to delve into the everyday micropolitics of how inhabitants negotiate their water access around existing structures to secure clean water. The following section describes how globalization has shaped cities as engines of growth, in the process leading to urban and peri-urban water insecurity and exacerbating related vulnerabilities. This section is followed by an empirical section based on fieldwork in Hyderabad and its surrounding peri-urban region; we describe the everyday problems of water scarcity in the city and its surroundings, privatization of water and the growing vulnerability of the inhabitants. This section is then followed by the conclusion.

5.2 Emerging Waterscapes in the Urban and Peri-Urban Global South

With the advent of modern globalisation, the global economy today is much more interlinked and interdependent than in the past. To tap into this massive network, cities have become the main platform for economic activities, and are becoming more entrepreneurial and outward-looking. They now have an intensifying need to reimage themselves as economic engines and magnets, in order to attract the jobs and investments required to attain the city's aspired economic development goals. As such, cities are in a race to brand themselves as "global", which is defined by Sassen (1991) as having a monopoly of command and control networks and institutions. According to Robinson (2002, p.534), a global city should "be able to articulate regional, national and international economies and serve as prominent nodes of a global economic system". Hence, globalisation is a key process that fuels a city's economic aspirations.

Since the mid-1990s, Hyderabad has been able to tap into these flows and emerged as a node in the global economy by attracting a large amount of foreign investments. This has successfully brought about rapid economic development and, along with it, an unprecedented growth in population through migration. However, in the bid to realise its economic and entrepreneurial aspirations to be recognised as an emerging global city, Hyderabad faces several major repercussions on its socio-spatial landscape. As Sassen (2005) reminds us, in many cases, cities that are highly integrated into the global economy are more likely to be disconnected from their surrounding areas rather than being able to stimulate territorial assimilation with their regional and national economies. The opening up of the economy and the expansion of Hyderabad have caused the neighbouring peri-urban villages to face many changes, such as large scale real estate development, decrease in agricultural land and increase in water stress (Prakash et al., 2011). Hence, Hyderabad's strategic location in the global network has caused an increase of inequalities between the city and its peripheral areas.

Peri-urban areas are a unique transitional space. Literature has conceptualised this rural-urban interface as a "space" in itself, consisting of highly dynamic interactions between the population and its landscape (Lerner & Eakin, 2011; Narain, 2010; Prakash et al., 2011). Many researchers have noted that peri-urban areas are frequently contested spaces, which have a growing heterogeneity of interests, and are therefore subjected to potential conflict (Douglas, 2006; Dupont, 2007). As claimants over key resources like land and water gradually multiply, the local natural resource base is impacted. This results in certain environmental and natural resource governance problems that urban or rural governmental or other institutions alone no longer have the ability to cope with (Narain, 2010; see also Lakshmisha et al., 2019).

Hyderabad offers land and basic facilities, such as free or subsidised water supply, to attract private investments and generate further growth (Prakash et al., 2011). As a consequence, the city is experiencing intensifying water shortage; its water table has dropped to such a level that water now has to be sourced from nearby areas outside its municipal boundaries (Prakash et al., 2015). This has brought about inequality in access to water for agriculture and other domestic uses in Hyderabad's peri-urban areas. Villagers are losing out on water access to the more powerful and well-off middle-class population residing in the urban core and to the powerful economic actors involved in peri-urban expansion. Moreover, urban expansion and encroachment into the surrounding transitional areas mostly takes place by encroachment upon forests, pasture commons and agricultural areas. As these are crucial for the natural water cycle, their conversion to urban infrastructure can disrupt this process, negatively affecting the recharging of the groundwater table (Prakash et al., 2011; Vij et al., 2019). Additionally, since these areas now host both agricultural and urban activities, the rise in number of users further contributes to the increase in water stress and insecurity for the villagers.

As the previous sections have established, there has been a depletion of water sources as a consequence of rapid expansion and urbanisation of Hyderabad; the public infrastructure is no longer able to keep pace with the growing needs of the city. Many scholars, including Prakash et al. (2011, 2015), Vij et al. (2019) and Lakshmisha et al. (2019), have documented that, as a result, a large informal groundwater market has emerged over time to bridge the growing gap between supply and demand. These informal water suppliers often get their water from sources in peri-urban regions, which are still richer in surface and groundwater.

Individuals living in the peri-urban areas who used to be farmers helped to fill this market gap by selling water to private tanker companies and small entrepreneurs from urban areas. With the onset of urbanisation, many former agriculturalists have turned away from farming and are taking advantage of the market conditions to sell water to private water tanker companies, who then drive to the water pumps in large trucks and transport the groundwater to customers and industries nearby or in the city (Prakash et al., 2015). This type of business constitutes an informal market, since neither the owners of the water pumps, nor the water tanker companies are registered with any government agency, and thus do not observe any formal guidelines (ibid.). Even though there are laws that regulate the construction of water pumps in most states of India, enforcement remains poor (Araral & Yu, 2013).

Allen et al. (2006) found that these tankers also supply water to peri-urban households, who have no choice but to pay market prices for water, since the formal supply is incapable of meeting their water needs. The authors distinguish between "policy-driven" and "needs-driven" practices in getting access to water services. They show that in peri-urban areas this access is primarily needs-driven, and that water is obtained through informal means rather than as the outcome of formal policies.

The growing number of users is continually increasing the stress on peri-urban water resources. Incessant exploitation has jeopardized the quantity of water available by negatively impacting the regenerative capacity of the groundwater systems. Water withdrawn far exceeds the amount that is recharged and, according to Prakash et al. (2015), depending on the amount of water that is being extracted, the Central Groundwater Board has listed the groundwater level of several sub-districts that adjoin the city of Hyderabad as critical and over-exploited. As such, groundwater depletion and water insecurity may be seen as both a cause and consequence of the emergence of the informal market; this being an extremely unsustainable cycle.

Rapid urbanisation accompanied by extensive water extraction from peri-urban areas has impacted the local residents' livelihood options. Prakash et al. (2015) observed that for some peri-urban households this has brought about new economic opportunities. There has been a shift in their sources of income, from mainly agricultural activities to more opportunities for employment in the city. As mentioned above, farmers are also taking advantage of the increase in demand for water by participating in the commercial sale of water through extraction of water from their bore-wells, which used to be for agriculture. Selling water to private tankers is a highly attractive option compared to agriculture: it requires little labour and investment but is much more profitable. These options help to diversify household incomes and thus provide households with greater economic security (Narain, 2010).

Vij et al. (2019) see the emergence of the informal water market as a process that has turned water into a tradable commodity, at the cost of the poor and marginalised villagers. Peri-urban areas are now increasingly catering to the water needs of the more well-to-do urban classes, a more prosperous clientele who have the ability to pay for a constant supply of water (Prakash, 2014). As a result, villagers are forced to purchase water at the same price. Prakash et al. (2015) observe that the private appropriation of what used to be common pool resources has led to intense local conflicts and water insecurity, as low-income groups have trouble paying for and thus obtaining water on a regular basis. The same research also found that a significant number of households receive less than the basic minimum standard amount of water, while tankers extract up to three times the amount that the village uses for daily household activities. Therefore, such flows of resources from peri-urban to urban areas may offer water security to city dwellers, at the expense of peri-urban residents, who are becoming highly vulnerable and water insecure (Vij et al., 2019).

Due to the informal nature of private water tankers, complaints from the villagers that their groundwater is being extracted for private gains are often to no avail. Most of these private companies operate their water tankers without any legal authorisation (Prakash et al., 2015). The main government body in charge of water issues in Hyderabad, the Hyderabad Metropolitan Water Supply and Sewerage Board, is only concerned with water supply within the city's municipal boundaries, and does not seem to pay any attention to the illegal operations in the peri-urban areas which are outside its purview (ibid). As such, these private tankers are not under the control of any authority and are free to operate as they like. Scholars have also documented that the farmers who are part of this illegal water trading business have the right to sell water as they wish, since groundwater access is tied to land-ownership through a legal and institutional framework that allows them to do so (Prakash et al., 2011, 2015). Villagers are thus unable to question such practices, which have even become more established due to the lack of a legal body to monitor their activities.

Research by the same authors has shown that peri-urban farmers participating in the commercial sale of water to urban communities are actually unaware of the adverse impacts this has on groundwater (ibid). To them, it is just an alternative livelihood and a way to deal with the dwindling agriculture and economic opportunities in the village. Nevertheless, they are still considered to be the luckier ones who prove resilient to the impacts of encroaching urbanisation and the appropriation of resources from the peri-urban areas. They are able to seize the opportunity for such a profitable activity whereas, the poor and marginalised peri-urban dwellers are left voiceless, vulnerable and deprived of their basic water supply.

The biophysical and chemical characteristics of water, and the quantification of water flows through the hydrologic cycle are often used as key indicators to determine the severity of water-related issues in Hyderabad. However, solely relying on these indicators as a determinant of these water-related issues treats water as a product of the hydrological cycle, postulating an undisrupted, endless circulation of water devoid of social meaning (Linton & Budds, 2014). Instead, urban political ecologists such as Swyngedouw (2004) and Loftus (2015) have underscored the inherently political nature of the production and distribution of water among the inhabitants of a city. Water itself embodies not only its biophysical and chemical characteristics (H_2O), but also contradictory and inseparable "social, economic, political and cultural processes" (Swyngedouw, 2004, p. 21) which are constantly reproduced over time as both a "*product* and *agent* of socio-natural change" (Lindon & Budds, 2014, p. 173, our emphasis). Swyngedouw's (2004) work on the social power of water illustrates this by revealing the political constituents of water that flows through the city. This water that flows is interwoven with the narrative of the state, financial capital, the production of rent, class relations, the process of water purification and the necessity of fulfilling an individual's physiological water needs. Moreover, access to clean urban water in some societies might also carry with it notions of "clarity, cleanliness, health and virginity" (Swyngedouw, 2004, p. 18). While we surface these complex entanglements, it is important that attention is also focused towards the scale of the body. By doing so, this would reveal how water-related inequalities are experienced by the inhabitants in their everyday life (Truelove, 2019). This would surface the power dynamics between various individuals and groups, and how this would alter their experiences surrounding water access and use (Sultana, 2009).

5.3 Methodology

Ethnographic fieldwork, when applied to research on water security, aims to surface how various socio-cultural and political processes related to water in the broadest sense are experienced, given meaning to and acted upon by people (see also Crang & Cook, 1995; Spradley, 1980). By doing so, it opens up opportunities to question and trace day-to-day actions and intentions exhibited by people, providing deeper insights on how they make sense of their surroundings (Herbert, 2000). Hence, through the knowledge co-constructed through photographs, interviews and the interactions with our participants, we attempted to interrogate and better understand the meanings and processes embroiled within water use, water access and water security in Hyderabad.

The fieldwork was conducted in December 2019 in Hyderabad, India. During this month, participant observations and semi-structured interviews were conducted in and around the city. The neighbourhoods of Madhapur and Nanakramguda were selected as they reflect the splintered landscapes of Hyderabad; resided by both locals and migrants, low to high income individuals and communities. While the participant observations were conducted along the streets of both neighbourhoods, young people aged 18 to 35 were approached through both random and snowball sampling, and subsequently semi-structured interviews were carried out. As these interviews progressed, some of our respondents allowed us access to their own private living quarters which gave us the opportunity to conduct further observations and discussions. Broadly, our interviews aimed to understand the socio-economic background of our respondents, their perspectives on water-related issues in Hyderabad and their day-to-day routines and interactions with the city's water resources. Subsequently, a thematic analysis on the verbatim interview transcription was done to identify key thematic strands and issues. These themes were triangulated with the field notes gathered from the participant observation to further develop on our analyses and identify nuances in our participants' interactions with water.

5.4 Uneven Landscapes: Differential Water Access and Use in and Around Hyderabad

5.4.1 Socio-Spatial Inequalities in the Built Environment

In transforming the urban landscape of Hyderabad, sprawls of land throughout the city were acquired by various state and private actors for various developments. These spaces, once acquired, are mostly rapidly transformed into various high-rise residential, mixed and office developments which are often occupied by the middle-upper social classes – spaces filled with *digerati*, white-collared office professionals and their families. These high-tech spaces are also premium infrastructural spaces which provide their inhabitants "elite" access to uninterrupted basic services like water supply. While these spaces provide "elite" access to facilities, peripheral spaces are bypassed, leaving behind the urban and rural poor (Das, 2015). While the adjacent peri-urban and rural spaces are now often known as sites of depleting groundwater supply, most urban spaces have the municipality to rely on to provide constant and clean water access to its inhabitants. Despite the municipality's intervention to ensure adequate water supply to these urban spaces, some inhabitants who live *within* these spaces still do not benefit from these facilities. Specifically, several water-related issues are still prevalent even within the urban built-up area of Cyberabad District.

Located within the financial district of Hyderabad, the neighbourhoods of Nanakramguda and Madhapur in Cyberabad are characterized by sprawls of high-rise residential developments and office buildings. Such developments, situated within these premium infrastructural spaces, are well-connected to other parts of the high-tech city through well-paved asphalt roads and concrete sidewalks, and also providing its inhabitants with the provision to uninterrupted access to basic services. While these might represent the modern high-tech business and financial hub that Cyberabad envisions them to be, these spaces are also dotted with landscapes representing the underbelly of the society. In the heart of Nanakramguda, spaces dotted with high-rise developments are lined with adjacent two to three-storey high concrete buildings, rows of zinc-roof houses and unpaved dirt paths in Nanakramguda (Figs. 5.1 and 5.2); a nod to the existing socio-spatial inequalities in Cyberabad. Hence, through our encounters with several inhabitants residing in Nanakramguda and Madhapur, we illustrate how these socio-spatial inequalities manifest through the access to and use of municipal piped tap water, and how its inhabitants negotiate their access to water through these structures.

Fig. 5.1 High-rise gated developments (left) and the adjacent neighbourhood (right) in Nanakramguda. (Photos by authors)

Fig. 5.2 Amazon's headquarters juxtaposed to a low-rise development in Nanakramguda. (Photo by authors)

5.4.2 Water as a Splintered Resource

Premium infrastructural spaces in Cyberabad are known to receive 24-h undisrupted municipal water supply. While connection to piped municipal water infrastructure in Cyberabad District is common, its supply is highly fragmented and disrupted. For households that do not receive municipal tap water, reliance on water from water tankers or bore wells is commonplace, while those who are linked to a piped water system might only receive water supply for a mere 1–2 h every alternate day. In our conversations with the inhabitants living in these areas, they reveal that spaces which are occupied by "professionals" are often provided with premium access. On the other hand, spaces occupied by inhabitants from native villages or holding low-skilled jobs often have to make do with disrupted or otherwise inadequate access to water supply. One migrant, who works as a shop assistant, notes this in our conversation: "What I've heard is that residential area[s], where VIPs (very important persons) used to stay, and in such areas, they get regular water [supply]. In the outskirts, village, low-class people, where they live, they experience water scarcity" (Interview with shop assistant, Hyderabad, 13 December 2019). The mentioning of VIPs points to the fact that these areas of "regular water supply" are commonly associated with people who are able to pay to reside in these spaces (e.g. high-rise gated communities). The spatial fragmentation of water access within an urban area illustrates how water is a highly splintered resource that is brought into the circulation of money and power (Swyngedouw, 2004). In this case, urban water networks in Hyderabad are tied with specific locations and types of accommodation, based on the commodification of water and on practices of granting or limiting access, depending on one's ability to pay for such services.

5.4.3 Exercising Agency in Attaining Water Access and Use

Water as a splintered resource in urban Hyderabad places restrictions on the inhabitants' water access. Despite this, inhabitants living in these spaces of limited water access do not succumb to these imposed structures. Rather, what has been observed is a form of individual and collective agency to find ways to secure their own water access and use. Exhibiting agency amid water crises is a common coping response to navigate the politics of water scarcity in the hopes of attaining water security. In poorer urban areas, Das and Skelton (2019) note that obtaining water from self-dug wells, borewells and private water tankers are common practices through which inhabitants try to secure access to water. Similarly, in their ethnographic study on peri-urban communities Narain and Singh (2017) note that inhabitants bypass state restrictions on extracting water from freshwater canals by installing handpumps in the vicinity, to benefit from the higher water table. By exercising various coping strategies, inhabitants circumvent existing structures in an attempt to be more water secure. However, the ways in which they cope with the inherent water insecurity are mediated through the intersections of social relations, and their needs and desires. While some inhabitants are able to exercise their agency to attain a more reliable and sustained clean water access and thus attain greater water security, others just find ways to cope in order to reduce the consequences of water insecurity.

In the neighbourhoods of Madhapur it is common for many households to only use the piped municipal tap water for non-consumption purposes (e.g. bathing or household chores). Water that is used for consumption is usually bought, with bottled 20-litre mineral water as a common option. The few households that decide to use tap water for consumption often purify the water, using a store-bought Reverse-Osmosis (RO) filter prior to consumption. Generally, households in Madhapur do not consume water directly from the tap. However, in contrast, most households in Nanakramguda use the piped municipal tap water for *both* consumption and non-consumption purposes. Only a few households boil the water prior to consumption.

While municipal water is piped to both neighbourhoods, the *ways* in which the water is used are different. A shop assistant living in Madhapur explains how water is used for households residing in Madhapur: "For drinking water, we will get mineral water from a shop, it is separate from the water which we use for bathing, washing, utensils. The drinking water is very much different, we use mineral water that we purchase". This is not only done in the shop keeper's household, but by many others who are residing in Madhapur. In our conversations, many have attributed this separation between consumable and non-consumable water use to concerns of piped water safety. As such, erring on the side of caution, they have avoided the (direct) consumption of tap water. A building engineer, for instance, aired his doubts about the piped water supply: "It is a safety thing, when the water from the tap is not good […] sometimes it will have some kind of bacteria. There might be chemicals in the water. In Hyderabad, they also mix some chlorine with water". These people have opted to consume store-bought mineral water to give them an assurance that the water consumed is safe. However, even while mineral water is a popular option in Madhapur, some even cast their doubts on its safety. A research assistant, who resides in a paying-guest residency in Madhapur, expressed her worries in our conversations: "I don't know if it is true, but [I have] heard that sometimes those mineral water bottles are not filled with very clean water. I don't want to fall ill, so it is better that I get an internal [RO filter] system." Water safety is not taken lightly in her instance, even if it means that she has to invest in a good RO filter. Our conversations show that many people have chosen to either purchase bottled mineral water or even purchase an RO filter as an additional safeguard to water safety.

Our conversations with various households in Nanakramguda have revealed a conception of water safety that differs from the situation in Madhapur. Although the households in Nanakramguda receive water from the same municipal supply of water, many households have indicated that the water is safe for direct consumption. However, to ensure that the water does not get contaminated, the households store drinking water in sealed containers or bottles and are kept away from the elements. Conversely, water that is not for consumption are stored in buckets or open water drums (Fig. 5.3). Some households, doubting the safety of tap water, also boil the

Fig. 5.3 Storage of potable and non-potable water in household. (Photos by authors)

water before consumption and do not purchase mineral water. When asked for the rationale behind their choice, they pointed out that, since the consumption does not result in any ailments, it is safe to drink. Moreover, since they are already paying Rs. 900 per month for the municipal water supply, they do not see the necessity of spending more just for mineral water. One resident of neighbouring Madhapur, empathises with the residents living in Nanakramguda on their choice: "For 10-litres of mineral water, it might take Rs. 30 to Rs. 40, which is costly, especially for a normal person, and the poor. They can't afford it, so they will drink the same water". With limited monetary resources, these inhabitants from lower social classes are often left with limited choices of mediating water insecurity.

While the discourse on water safety largely differs in both neighbourhoods, our conversations with the locals and the ground realities of water use have revealed the fractured nature of water use within Cyberabad. This fracture is largely seen across various social classes and, in particular, the ability of individuals or households to afford mineral or RO-filtered water. The resident of Madhapur mentioned above said that her husband's and her own safety should not be compromised, even if it means paying *more* for the assurance of clean water: "obviously [consuming RO-filtered water] is incurring more money but safety is something that [I] will not compromise [on]. Since water is scarce in Hyderabad and part of our daily needs, spending a little [more] is alright". As a comparison, the installation of an RO-filter costs an average of Rs. 4500, excluding the costs of maintenance and filter membrane changes amounting to an additional of Rs. 2000 per year. However, mineral water only costs an average of Rs. 15 to Rs. 40 per 20 litres, while the use of tap water does not involve additional costs. The choice to pay more to ensure water safety reflects how water is socially stratified and embroiled within the water/money nexus (Swyngedouw, 2004).

The realities in both neighbourhoods showed that many inhabitants have demonstrated various ways of exercising agency to cope with the lack of *clean* water supply. However, what is socially stratified are the *ways* in which the inhabitants cope.

These ways reflected in the inhabitants day-to-day life, where those who are able to pay for such affordances look towards more formal means (i.e. purchasing mineral water and using an RO-filter), while others, towards more informal means (i.e. covered storage and boiling of water).

5.4.4 Securing Water Access and Gender-Water Relations

In urban Hyderabad, intermittent water supply through various neighbourhoods is common. In order for the inhabitants to secure their access to a continual supply of water, many of them have looked to drawing water from borewells, water tankers or storing water from taps in containers (Das & Skelton, 2019). While the inhabitants have found various ways to cope with the lack of (clean) water supply, our conversations have also revealed that the women in the households, often the wife or the daughter, are often the ones bearing the burden by being the primary "caretaker" of water collection. As such, these women would need to structure their time against water collection timings, bear the physical and mental load of water collection to ensure that the household is water secure. Consequently, the men, unless they live alone, are usually not involved in these duties, as their primary responsibility is to earn an income for the family. However, women, particularly from the lower social classes, are often working hard as well to provide for the family or pay for the education of children, next to their water-related and other household duties.

In one of our interactions with a 19-year old female high-school graduate who resides in Nanakramguda, she shares that she is "responsible" for water collection for her family of three. As the youngest, who is neither working nor studying, she shoulders this responsibility by waking up at six in the morning to store water for her family. In situations where the municipal pipeline ruptures, cutting off water access, she recalls that she would be responsible to obtain water for her family. This is done by either purchasing mineral water or obtaining water from water tankers to ensure sufficient water. Accounts like these, where the woman of the household shoulders water collection duties are commonplace among lower-income families in Hyderabad. Similar to other studies, such as Das and Safini (2018), the gendered division of labour and dependence on their male counterparts to foot for water bills have left women more vulnerable and water insecure in Hyderabad.

While women often bear the brunt of water collection duties in this neighbourhood, an individual's social class can further complicate gender-water relations. In conversations with other women residing in Madhapur, we found that there are two situations where they are not responsible for such duties. In the first case, women do not involve themselves in water collection duties if the household is located within a neighbourhood where there is an undisrupted and clean supply of water. In another case, if the household is located in an area with an interrupted water supply, households with helpers/servants will be responsible for water collection duties. A woman, for instance, living in a family of five explains her household's water collection routine: "My servant will be responsible for storing the water, and sometimes, my father who is retired will help her. But, my mother, brother and me are working, so we do not do it" (Interview with secretary, Hyderabad, 13 December 2019).

These two cases further demonstrate how water is embroiled with the notion of power. Whether living in a "premium infrastructural space" with a constant clean water supply or a household with a helper, the ability for a household to pay for such affordances leaves the women of the household free from water collection duties. However, the helper who assumes the water collection duties tends to be a female. The involvement of a *female* helper is a nod to how class differences are further entwined with gender-water relations. It is also a testament to how these embodied gendered experiences of securing water access not only differs between different households, but also *within* a household.

5.5   Conclusion

The rapid urban expansion in Hyderabad has vastly transformed its landscape to a high-tech metropolis supported by a robust infrastructure system. While these premium infrastructural spaces enjoy the affordances brought by robust state-initiated infrastructure, expansion of the urban community has been at the expense of part of the population residing in the periphery. These marginal peri-urban and urban communities are often bypassed by public services and facilities, suffering from depleting water resources and strained water access. Hence, water in Hyderabad is a highly fractured resource across the intersections of social class and gender.

Still presented as an engine of growth for the region, Hyderabad is not near the end of its conquest for urban expansion. This means that depletion of water resources, privatization of water, splintering of water infrastructure, water contamination and a slew of other water-related issues will probably continue and even exacerbate. However, while the inhabitants of Hyderabad have demonstrated various ways to mediate their access to water, what needs to be brought forth are how these acts, through formal and informal means, remain heavily splintered across social classes. The "water-money nexus" enables inhabitants from middle-upper social classes to attain water security by successfully negotiating their water access amid Hyderabad's water crises, thus creating greater water security for themselves. On the other hand, those inhabitants who are not able to engage in or afford such negotiations can only cope through informal means that often provide less secure access to water, leaving them at an even more precarious position of water insecurity.

Inhabitants themselves are, to a greater or lesser extent, capable of exercising agency to cope in various ways with the lack of access to clean water. Despite this being so, only some are able to successfully negotiate their access to water; hence, many remain water insecure. In the face of rapid urbanisation and exacerbated water-related insecurities, can peri-urban communities find ways to sustainably attain a higher degree of water security for all?

Acknowledgements We would like to acknowledge the funding support from Nanyang Technological University - URECA Undergraduate Research Programme (IRB-2019-11-022) and NIE AcRF RI 1/17 DKD for this research. We would also like to thank our fieldwork assistants in Hyderabad as well as our NIE Geography BA batchmates (Class of 2020) that helped us during the fieldwork.

References

Allen, A., Dávila, J. D., & Hofmann, P. (2006). *Governance of water and sanitation Services for the peri-urban poor. A framework for understanding and action in metropolitan regions.* Development Planning Unit, UCL.

Araral, E., & Yu, D. J. (2013). Comparative law, policies, and administration in Asia: Evidence from 17 countries. *Water Resources Research, 49,* 5307–5316.

Crang, M., & Cook, I. (1995). *Doing ethnographies.* Geobooks.

Das, D. (2015). Hyderabad: Visioning, restructuring and making of a high-tech city. *Cities, 43,* 48–58.

Das, D., & Safini, H. (2018). Water insecurity in urban India: Looking through a gendered lens on everyday urban living. *Environment and Urbanisation ASIA, 9*(2), 178–197.

Das, D., & Skelton, T. (2019). Hydrating Hyderabad: Rapid urbanisation, water scarcity and the difficulties and possibilities of human flourishing. *Urban Studies,* 1–17.

Dupont, V. (2007). Conflicting stakes and governance in the peripheries of large Indian metropolises–An introduction. *Cities, 24,* 89–94.

Douglas, I. (2006). Peri-urban ecosystem and societies: Transitional zones and contrasting values. In D. McGregor, D. Simon, & D. Thomson (Eds.), *The peri-urban interface* (pp. 18–27). Earthscan.

Herbert, S. (2000). For ethnography. *Progress in Human Geography, 24*(4), 550–568.

Lakshmisha, A., Priyanka, A., & Manasi, N. (2019). Assessing the double injustice of climate change and urbanization on water security in Peri-urban areas: Creating citizen-centric scenarios. In *Water insecurity and sanitation in Asia* (pp. 301–323). Asian Development Bank Institute.

Lerner, A. M., & Eakin, H. (2011, December). An obsolete dichotomy? Rethinking the rural–urban interface in terms of food security and production in the global south. *The Geographical Journal, 177*(4), 311–320.

Linton, J., & Budds, J. (2014). The hydrosocial cycle: Defining and mobilizing a relational-dialectical approach to water. *Geoforum, 57,* 170–180.

Loftus, A. (2015). Rethinking political ecologies. *Third World Quarterly, 30*(5), 953–968.

Narain, V. (2010). *Peri-urban water security in a context of urbanization and climate change* (Peri urban water security discussion paper series, discussion paper no 1: SaciWATERs). Retrieved from http://saciwaters.org/periurban/idrc%20periurban%20report.pdf

Narain, V., & Singh, A. K. (2017). Flowing against the current: The socio-technical mediation of water (in)security in periurban Gurgaon, India. *Geoforum, 81,* 66–75.

Ramachandraiah, C., & Bawa, V. K. (2000). Hyderabad in the changing political economy. *Journal of Contemporary Asia, 30*(4), 562–574.

Prakash, A., Singh, S. and Narain, V. (2011). *Changing waterscapes in the periphery: Understanding peri-urban water security in urbanising India.* Oxford University Press.

Prakash, A. (2014). The peri-urban water security problem: A case study of Hyderabad in Southern India. *Water Policy, 16,* 454–469.

Prakash, A., Singh, S., & Brouwer, L. (2015). Water transfer from Peri-urban to urban areas: Conflict over water for Hyderabad City in South India. *Environment and Urbanisation ASIA, 6,* 41–58.

Robinson, J. (2002). Global and world cities: A view from off the map. *International Journal of Urban and Regional Research, 26*, 531–554.

Sassen, S. (1991). *The global city: New York, London, Tokyo*. Princeton University Press.

Sassen, S. (2005). *The global city: Introducing a concept*. Brown Journal of World Affairs.

Sen, S. (2017). Peri urban water woes and development contradictions. *Geography and You, 101*, 74–79.

Spradley, J. P. (1980). *Participant observation*. Harcourt Brace Jovanovich College Publishers.

Sultana, F. (2009). Fluid lives: Subjectivities, gender and water in rural Bangladesh. *Gender, Space and Culture, 16*(4), 427–444.

Swyngedouw, E. (2004). *Social power and the urbanisation of water: Flows of power*. Oxford University Press.

Truelove, Y. (2019). Rethinking water insecurity, inequality and infrastructure through an embodied urban political ecology. *WIREs Water, 6*(3), 1–7.

Vij, S., John, A., & Barua, A. (2019). *Whose water? Whose profits? The role of informal water markets in groundwater depletion in peri-urban*. Hyderabad.

Chapter 6
Changing Agriculture and Climate Variability in Peri-Urban Gurugram, India

Pratik Mishra and Sumit Vij

6.1   Introduction

A recent essay on climate change adaptation (CCA) mentions an instance of urban flooding that highlights the stakes of uneven climate protection regimes (Yarina, 2018). In the 2011 flooding of the Chao Phraya delta in Thailand that killed over 800 people, the King's dyke protecting the Bangkok metropolitan area kept the capital dry, while the displaced floodwaters made conditions in peripheral districts worse. Protesting inhabitants from these districts descended on the protected areas, opening the floodgates and tearing holes in the sandbag walls, while the prime minister counselled them to think of the national good. If the city centre flooded, she said, it would cause "foreigners to lose confidence in us and wonder why we cannot save our own capital" (Yarina, 2018, p. 1). The example asks us to urgently and critically consider the way our cities are constructed: city peripheries or peri-urban spaces often exist in an exploitative relationship with the urban, and this inequity could be further exacerbated by the effects of climate change and climate change adaptation (CCA) policies.

In India, a new wave of farmers' movements has taken centre stage, as thousands of farmers marching on cities have captured the public attention and forced us to acknowledge what should have been apparent from the epidemic proportions of farmer suicides every year: that there is a deep structural crisis in Indian agriculture. Yadav (2017) states that the crisis is three-fold: economic, ecological and existential. The economic crisis comprises the volatility of market prices, the stagnation of

P. Mishra (✉)
Department of Geography, King's College London, London, UK
e-mail: pratik.mishra@kcl.ac.uk

S. Vij
Department of Geography, King's College London, London, UK

Public Administration and Policy Group, Wageningen University & Research,
Vrije University, Amsterdam, Netherlands

V. Narain, D. Roth (eds.), *Water Security, Conflict and Cooperation in Peri-Urban South Asia*, https://doi.org/10.1007/978-3-030-79035-6_6

farmer incomes and a lack of capital for modernization. The ecological crisis in Indian agriculture is deeply associated with the after-effects of the Green Revolution: the insatiable treadmill of higher fertilizer, pesticide use and deeper borewells, rising incidence of cancer from pesticide use, and the metaphorical cancer of indebtedness of farmers, swiftly progressing from being manageable to becoming debilitating. These two crises culminate in the existential crisis of Indian agriculture, marked by farmer suicides, distress migration out of farming, and the unravelling of agrarian futures from climate change impacts.

This chapter locates the changing dynamics of agriculture at a frontier where a geographically specific articulation of this crisis comes to the fore: in the peri-urban villages bordering Gurugram City.[1] These villages are still largely agrarian but undergoing rapid and radical changes. A major characteristic of peri-urbanization are the land and water use changes caused by urbanization, such as the speculative buying of farmlands and their transformation into other (non-agricultural; e.g. services, industry) uses, construction of farmhouses, and the siting of canals and water and sewage treatment plants that service the city. In this chapter, the peri-urban areas are understood as dynamic spaces emerging due to rapid urbanization, where agricultural land and water uses are continuously changing and the underlying social, economic, political and institutional systems are in a state of flux (Vij et al., 2018). The core emphasis of our research is to investigate the double exposure of Gurugram's peri-urban space to both urbanization and impacts of climate variability upon peri-urban agriculture. Our chapter looks particularly at how this spatial inequality plays out in peri-urban Gurugram and how climate variability is gradually unfolding and culminating into differential impacts for peri-urban agriculture. Climate variability is primarily represented here by the unpredictability in rainfall patterns, which is increasingly affecting farming and related livelihood activities.

The remainder of this chapter is divided into three sections. In Sect. 6.2 we explain why the peri-urban should be considered a relevant object for analysis. Most analyses of climate change and vulnerability tend to select either rural or urban contexts (Narain & Prakash, 2016, p. 2). We take forward the emerging research agenda of focusing on a peri-urban context of vulnerability and CCA that puts climate change in conversation with the urbanization literature. It details the specific socio-natures produced in peri-urban spaces that villagers, especially farmers, must contend with (Vij et al., 2018). Section 6.3 presents the key findings, explaining the effects of climate change upon peri-urban agriculture. Section 6.4 wraps up the arguments, provides some policy recommendations, and sketches priorities for future research.

6.2 Researching the Peri-Urban

Peri-urban spaces fall within the gap between the urban and rural CCA literature, which emphasise built environments or agrarian systems respectively. The peri-urban, as a space where rural and urban land uses co-exist (Bowyer-Bower, 2006)

---

[1] Gurugram was called Gurgaon until 2016.

or, more broadly, as processes of transition between rural and urban with adverse environmental impacts (Allen, 2003; Narain & Nischal, 2007; Vij et al., 2018), raises specific questions related to its transitional characteristics. There is a growing interest to understand the peri-urban analytically, as a dynamic space where adaptive strategies and mediating institutions are couched within logics of incipient urbanization or the co-existence of agrarian and urban features in a rapidly changing landscape. We consider peri-urban spaces as a relevant analytical category for two reasons.

Firstly, peri-urban spaces exist not only at a spatial intersection between the rural and urban, but also at a temporal intersection of the anticipation of urbanization, where the future might be radically unmoored from exigencies of the present, and its extended present, rocked by the currents of the Indian agrarian crisis. As peri-urban spaces are in a state of institutional flux, CCA measures may be extremely difficult to realize there. This is because planned CCA measures are generally implemented by the state (Preston et al., 2011). Peri-urban challenges are hardly recognized in policy circles and, moreover, the peri-urban is currently missing or not prioritized in newer policy domains such as climate policy (Marshall et al., 2009; Roth et al., 2019; Vij, 2019). As CCA focuses on different temporal (short, intermediary and long-term) and spatial (rural, urban, coastal) scales, the institutional flux in peri-urban spaces makes it challenging to fit the peri-urban into a policy framework and to implement that on the ground (Eakin et al., 2010). Further, the state follows a conventional sectoral and structural interpretation of climate impacts, mostly governed by its agencies responsible for water and agriculture. The current system does not take into account the specific impacts of climate change in peri-urban areas and has limited understanding of the ways to address the role of urbanization in these spaces, which creates a policy and institutional gap (Allen, 2003).

Peri-urban spaces mark sites where the material production of the urban is most intensely pursued as well as contested by various state and non-state actors. Peri-urban developments have often been executed by conceiving of these areas as *tabula rasa* or "blank slate", upon which the financial and political might of the city can enact its vision of urban expansion, literally bulldozing over local agrarian histories and meanings of place. In looking at "ordered" spaces of global capital created in the periphery, Ong (2006, p. 19) remarks that "nowhere are logics of […] spatial purification more readily deployed than in the creation of new urban spaces of intensified neo-liberal exceptions on the urban periphery". "Spaces of exception", like Special Economic Zones or large public and private projects (airports, universities, IT parks), often pushed through without social or environmental auditing or proper democratic processes, are usually located at the peri-urban fringe of major cities, and positioned as modernization and development.

The peri-urban is important in relation to the ongoing urbanization of spaces and therefore for low-income migrant workers who prefer to live in close proximity to their work sites (Kundu, 2009; SELCO Foundation, 2016). The regulation of land use is more relaxed than in urban areas, so migrants find it easier to reside here (Sridhar, 2010). Peri-urban areas, then, become an important hub for migration of people out of rural areas for work. Physical migration, both seasonal and permanent, from rural areas has intensified in India in the last two decades. Most migration into urban areas is not a move into the modern sector: 94% of rural-urban migrants enter the informal sector, 60% of which are in the self-employment category (Murthy, 2013). In other words, workers end up in the urban informal sector as low-wage workers or self-employed self-exploiting petty commodity producers (ibid.). This migration is a symptom of the shedding of labour from the agrarian sector into urban manufacturing and service sector jobs, often with no better prospects. Much of this migration is forced, and results from the inability of wage workers to survive on smallholdings in agriculture (Colatei & Harriss-White, 2004). In the context of the agrarian crisis, Dipankar Gupta (2005, p. 757) scathingly writes:

> Agriculture is an economic residue that generously accommodates non-achievers resigned to a life of sad satisfaction. The villager is as bloodless as the rural economy is lifeless. From rich to poor, the trend is to leave the village [....] The town is not coming to the country, as much as the country is reaching out to the town.

Migration between the village and the city in India has become a highly unequal, spatially uneven relationship, determined and restructured by the general situation of agrarian crisis. This crisis, which is ubiquitous in Indian agriculture, is reproduced and intensified in peri-urban villages as well, but constituted through several contingent factors operative on local, regional, national and global scales. Metabolic geographies of peri-urban areas, as we present in this chapter, are the product of this context of agrarian decline in ways not immediately apparent: the devalorisation of farming and the cynical outlook on agrarian futures among farmers emerge from this distress and influence possible futures and CCA strategies in the peri-urban space, where greater affinity towards the city is the norm.

The second important reason for taking the peri-urban as an analytical category concerns the rich insights it offers us in relation to processes of urban metabolism. The burgeoning field of Urban Political Ecology (UPE) has provided rich insights in how cities are constructed out of the flows of urbanized nature and has done much to articulate the spatially distanced injustice engendered in flows that place the richest areas and citizens in an exploitative relation with the poorest (see Furlong & Kooy, 2017; Ranganathan & Balazs, 2015; Swyngedouw, 1997). The peri-urban is an active yet under-studied site of urban metabolism. It has long been the bearer of infrastructures for the urban that, for environmental and economic reasons, are relocated there. Usually these infrastructures, which service the city's needs (metabolic or otherwise), produce few benefits to their peri-urban surroundings. In his historical study of Delhi, Sharan (2014, p. 18) mentions how infrastructures and industries that were seen as "nuisance" or "noxious" were moved outside the urban area by both colonial and post-colonial administrators. In this chapter, we focus on metabolic infrastructures, specifically canals, that have been constructed in or through the peri-urban, on the basis of similar logics of outsourcing from the city into devalorized space.

These relations of producing the urban and of serving urban metabolic needs by being sites and conduits for urban infrastructure make the peri-urban a productive analytical lens for thinking through urbanization, also in relation to climate change and CCA. We see the term "peri-urban" as complementary to a range of concepts, categories and analytical lenses that help us better understand the framings and spatial scales at which different processes can be addressed.

6.3   Context and Methodology

6.3.1   *Budhera*

The authors conducted their research in Budhera, a village on the outskirts of Gurugram city (see Fig. 6.1). The concentration of urban-oriented metabolic infrastructures predominates, thus putting the stamp of urbanization upon it. Although it is also rapidly urbanizing in other respects, Budhera's peri-urbanity could be considered synonymous with being saturated with the markers of urban metabolism. The village is a site to two drinking water canals, one wastewater canal, and a large water treatment plant, besides being bordered by the Delhi region's most important sewage channel, the Najafgarh drain, and criss-crossed by multiple high-voltage power lines. Budhera's landscape is marked throughout by infrastructures that service urban metabolism. The village landscape is a mix of agricultural farmland, urban-support infrastructure and farm-houses of the urban elite. The villagers

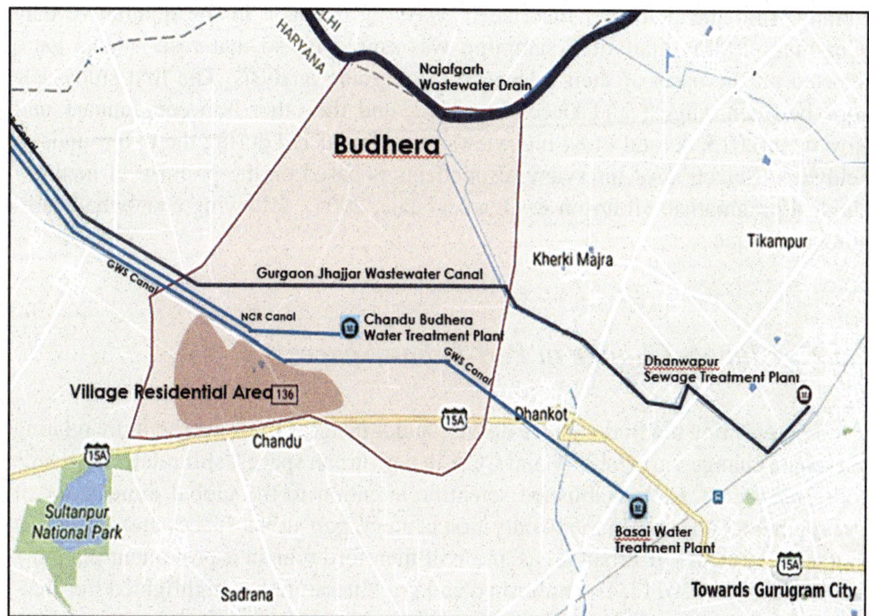

Fig. 6.1 Map of Budhera and its surroundings, marking the canals and the water treatment plant. (Source: authors, using Google maps)

maintain close ties with the city of Gurugram for their consumption, education and medical needs, through jobs in the urban economy and for marketing of their agricultural produce.

Since the economic reforms of 1991, the state government of Haryana has harnessed the growth potential of Gurugram. Its population stood at 1.5 million in 2011, against 0.8 million in 2001 (Census, 2001, 2011).

The rapid growth of Gurugram is due to its proximity to Delhi and the domestic and international airport, and to the policies pursued by the state government to attract multi-national companies and real estate business (Narain, 2014; Vij et al., 2018; Vij & Narain, 2016). The rapid growth of the city that has in the past subsumed other peripheral villages in its urban spread, provides the backdrop to a state of anticipatory urbanization in Budhera. Anticipatory knowledge on urbanization draws attention to a possible future and reflects on responsive measures for the present by peri-urban residents (White, 2016). Villagers expect high-rise buildings and shopping malls to arrive at their doorstep within the coming decade, rendering obsolete their age-old occupation of farming. This anticipation was expressed by a farmer stating that "in four years, I will sell this land to a builder, get a flat and open a grocery shop in the area, sitting on a chair all day with no more work to do".

This chapter is based on qualitative research by each of the authors. Both authors followed a case study design, mostly using semi-structured interviews but also methods of participant observation, transect walks and focus group discussions. An empirico-inductive framework was used so as to be reflexive to issues and findings coming from the field and theorize in ways "grounded" in the qualitative data (Charmaz, 2006). Analytical sampling was employed so that respondents were selected on the basis of their relevance to on-going analysis. The first study was done between August and December 2012, and the other between January and November 2015. A total of 84 interviews were conducted during the two rounds of fieldwork. Selection of interview respondents is based on the premise of methodological pragmatism (Johnson & Onwuegbuzie, 2004), following a snowball sampling technique.

6.3.2 Climate Change in Peri-Urban Spaces

Before presenting the findings, we discuss major insights from the literature relating to climate change vulnerability and CCA in peri-urban spaces, still relatively scarce compared to the well-established scientific attention to the global dimensions of environmental change. In the twenty-first century, peri-urban spaces are growing at four times the rate of urban areas and will therefore remain a prominent planning challenge (Piorr, 2011). The published and grey literature has highlighted the challenges of peri-urban agriculture in urbanizing regions, largely focusing on agriculture from the perspective of livelihoods, poverty reduction and ecosystem services. However, little attention has been paid to explaining the interlinkages between socio-economic and political changes, the vulnerabilities of marginalized communities and climate change variability (Lwasa et al., 2014). Much literature focuses on formal policies and interventions, in which state and non-state actors are tasked to implement National Adaptation Programmes of Action (NAPAs[2]), National Adaptation Plans (NAPs[3]) and other climate policies through stronger local investment and action in developing countries. However, relatively little attention is paid to the new organizational and institutional arrangements developing locally in day-to-day social interactions and practices.

Further, peri-urban populations disproportionately receive the negative externalities of urban growth and changing climate without the associated benefits of services and assets (Allen, 2003; Allen et al., 2017). With these challenges in peri-urban spaces, CCA strategies are necessary. CCA scholars argue that there are barriers to CCA, especially social, economic and political ones. Moreover, peri-urban impacts of climate change are not clearly explained with reference to the lack of institutional capacities and uncertain knowledge of climate change in peri-urban spaces. This can be partly linked to the lack of popularity of the concept of the peri-urban in the literature and its complexity in terms of spatial dynamics. The peri-urban can rather be seen as a new kind of highly dynamic and multi-functional territory, where there is inadequate spatial governance (Ravetz et al., 2013).

With rapid urbanization, there is an increase in public and private encroachment of common pool resources (CPRs) such as rainwater harvesting structures and cumulative variability in rainfall and temperature is reducing the dependency of communities on CPRs (Vij & Narain, 2016). Ranjan and Narain (2012) note that peri-urban residents suggest that decreasing rainfall is the cause of reduced dependence on traditional rainwater harvesting structures in peri-urban Gurgaon. With very little water in these structures, the investments needed to maintain these structures are a challenge. Such CPRs in peri-urban spaces were traditionally utilized to support earthen dams and infiltration ponds on hill slopes to slow down runoff and possible flooding. Further, the rainwater harvesting structures were also providing water for year-round production of urban and peri-urban crops. In many peri-urban villages, the adaptation measures were in place historically by means of CPRs that dealt with water scarcity. However, with depleting and encroached CPRs, it will be partly challenging to protect communities from variability of climate change.

---

[2]NAPA is a process for Least Developed Countries (LDCs) to identify priority activities that respond to urgent needs to adapt to climate change; see https://unfccc.int/topics/resilience/work-streams/national-adaptation-programmes-of-action/introduction

[3]The NAP process was established under the Cancun Adaptation Framework (UNFCCC-CAF, 2010). It enables countries to formulate and implement national plans that identify medium- and long-term adaptation needs. NAPs further help in developing and implementing strategies and programmes to address those needs. It is a continuous, progressive and iterative process which follows a country-driven, gender-sensitive and participatory approach.

6.4   Commons, Canals and Peri-Urban Agriculture

In this section we use two processes associated with urbanization that are observable and salient in the peri-urban space to create a situated, differentiated understanding of the combined effects of urbanization and climate change-related stress. These processes are, first, the current and future loss of CPRs to urban expansion and, second, the emergence of drinking water and wastewater canals as an expression of urban metabolism, disrupting the peri-urban landscape with new externalities but also creating new opportunities.

CPRs are a category of natural resources that are owned, controlled and managed by a group of people. It includes community pastures, ponds, forests, and wastelands — all important natural resource endowments in the rural areas of India (Jodha, 1986). CPRs especially support the livelihoods of marginal and landless households that have little alternative means in terms of private assets. Ostrom (1992) has contributed to a scientific understanding of the conditions under which CPRs can be managed sustainably and thus survive over generations. Since the early 1960s, CPRs have been declining in India (Ballabh et al., 2002; Jodha, 1985; Manikandan & Sundaram, 2017). In India this process is caused by several factors, such as privatization of common resources, commercialization and modernization of agriculture, population growth and land reform programmes. Narain (2014), Rakodi (1999), Vij (2014), and Vij and Narain et al. (2016) have pointed out that CPRs are declining, and even more rapidly so in peri-urban spaces. Thus, the latter two publications have shown how these intensified processes in peri-urban spaces are related to the increase in competition for resource uses between urban, peri-urban, and rural resource-dependent populations. In the following sections, we present findings from our research that speak to this concern.

A major focus in our research was the emergence in a single village of multiple infrastructures that support urban metabolism. We identified two drinking water canals —the Gurgaon Water Supply canal (GWS canal; see Fig. 6.2) and the National Capital Region canal (NCR canal)— and two wastewater channels —the Gurgaon-Jhajjar Wastewater canal (GJC Canal or wastewater canal) and the larger Najafgarh Drain— that all cross the boundaries of Budhera village. Further, the Chandu-Budhera Water Treatment Plant occupies 274 acres of land that have been taken from the agricultural and common grazing lands of the village. In Budhera, they are as much an expression of urbanization as the residential apartments, shopping malls and recreation centres that form the more conventional imaginaries of urbanization.

While CPRs such as grazing land and traditional rainwater harvesting structures in Budhera are degrading or decreasing, these various canals are becoming part of the peri-urban waterscape (Vij et al., 2018; Vij & Narain, 2016). We consider canals as emerging CPRs in peri-urban areas, as these structures are managed and governed by agriculture-related norms in a way that is, to some extent, similar to those that have historically developed in the disappearing CPRs (Vij & Narain, 2016). For instance, when a wastewater canal starts to be used for irrigation, farmers engage in

Fig. 6.2 The GWS drinking water canal and the newer NCR canal to the left running parallel alongside the fields. (Photo Sumit Vij)

forms of collective action, devise rules on the use and management of the water, and thus are (re-)defining the canal and its water as a form of property (von Benda-Beckmann et al., 2006), as "hydraulic property" (Coward Jr & Levine, 1987). In these processes, such emerging infrastructures become the sites of new forms of cooperation and conflicts between peri-urban residents (Vij et al., 2018).

The impact that these infrastructures have on the lives and livelihoods of villagers in Budhera are significant. The canals and treatment plants behave as open, leaky systems and, through the processes of seepage, irrigation, breaches and groundwater recharge, become part of the social and hydraulic landscape of Budhera as much as they are part of Gurugram city. Some of the insights from the study were recently published in a paper that sought to establish the presence of metabolic infrastructures as an important constituent when conceptualizing the peri-urban fringe (Mishra & Narain, 2018).

An instance of this concerns the GJC, which is used by farmers in Budhera as a source of nominally priced (almost free) irrigation water (Fig. 6.3). This canal conveys treated discharge from the Dhanwapur sewage treatment plant (one of Gurugram's two major STP plants), passing through Budhera and several other villages before merging into the larger Najafgarh Drain (Delhi region's major wastewater outlet). The practice of irrigation from the GJC is state-sanctioned, against a nominal charge of INR 50 (or $0.72) per year imposed on farmers for use of wastewater.

Fig. 6.3 Agricultural fields in Budhera. A motor pumps water from the wastewater canal. (Photo Pratik Mishra)

Wastewater not only irrigates the fields but also partially fertilizes it, reducing input costs for farmers. The wastewater canal flowing through Budhera is bunded at a higher level than the adjacent fields, so irrigating from it is merely a question of opening the seal from pre-installed outlets in the canals. Wastewater is thus (almost) free and quick (the 12-inch pipe outlets take only 3–4 h to irrigate an acre) and can be widely distributed through furrows that extend up to a few kilometres from the canal. Thus, it irrigates fields located farther off, maintained by an informal cooperative arrangement among farmers. There are, however, problems with wastewater irrigation as well. Once there was a tragic mix-up of industrial effluents from another canal into the GJC. Farmers, unwitting of this danger, irrigated their fields with heavy metals that settled in the soil and brought down productivity drastically. Farmers also have an aversion against irrigating food with wastewater, which they consider "dirty". They don't consume any produce from their own harvests of wheat and rice, when grown with wastewater. But nevertheless, wastewater irrigation is a popular technology that is used by most farmers in Budhera.

Being so close to Gurugram, a large number of young people and laborers from Budhera have moved into the city or commute for work. There have been many speculative purchases of land here in the preceding decades, mainly by outsiders who had identified the potential for financial gains from rising land prices. The absentee landlords of both these categories lend out their land on so-called *kannbataai* (sharecropping) arrangements to tenants. In this system the tenant either makes an upfront payment to the landowner when entering the agreement, or pays a fixed rent of grain and fodder after harvest. Both the *kann-bataai* tenants and the landowners tend to be short-sighted on the agricultural productivity of land and find wastewater irrigation cost-effective. Hence they were among the early adopters of wastewater. Because, according to farmers, crops grown with groundwater irrigation tend to spoil or rot when wastewater enters into the mix, farmers of land adjacent to the wastewater canal are forced to cultivate using its water. This domino effect means that most farmers in Budhera cultivate through the GJC wastewater canal.

In Budhera, it has been noted that the rising demand for land and the increasing pressure of population have led to encroachment of water bodies called *johads* (earthen rainwater harvesting structures or ponds). Moreover, with higher rainfall variability, the lowering of the water table has pushed farmers in Budhera to use ever more costly tubewells to be able to access water for irrigation (Vij, 2014). The high costs of digging deeper tubewells has deprived marginal farmers of access to groundwater, eventually bringing changes in the occupation of marginalized peri-urban communities (Vij & Narain, 2016). Respondents in Budhera repeatedly mentioned this changing availability of water for some as a result of changing patterns of rainfall and growing stress on groundwater. One of them mentioned that Budhera used to be famous for its cultivation of muskmelon, and that families in the village used to receive marriage proposals from elsewhere thanks to the availability of fresh water and muskmelon. With the decline in rainfall and growing pressures on groundwater from the surrounding areas, the cultivation of muskmelon has completely stopped.

6.5 Changing Agricultural Patterns with Climate Variability

The main manifestations of climate change in Budhera are the increasing unpredictability of weather cycles and extreme events, and the uncertainties this is creating for farmers cultivating their crops with wastewater. Studies in and around Budhera reveal several ways in which a changing climate is experienced by the residents (Narain & Singh, 2017; Ranjan & Narain, 2012). These include a longer duration of summers, shorter winters and changes in patterns of precipitation. Many local residents report a disappearance of the *chaumaasa*, or the 4 month monsoon period of rainfall. These observations are also corroborated by the analysis of hydro-meteorological data (see Narain et al., 2016). When this research was conducted in 2015, three out of the four previous harvests of rice and wheat (the crops in the *kharif* and *rabi*[4] seasons respectively) had resulted in an economic loss for farmers, twice from unseasonal rains and pests, and once from low market prices. In March 2015, unseasonal rains struck Budhera and damaged the winter wheat harvest that was near to being cut. This period was recorded as the wettest March in 48 years.

---

[4] The monsoon and winter crops.

In February 2015, Caftanji, a farmer had already been worried about watering his fields using wastewater. Three days of strong sunshine, unusual for chilly February, had begun to wilt the wheat. He irrigated his field, but when the weather reverted to the cold temperatures typical of February, the wastewater standing in the field damaged crop and soil. This problem is more acute than with groundwater because wastewater takes longer to evaporate. The situation was worsened by seepage from the drinking water canals that had raised the water table in adjoining fields. Thus, timing of the watering cycles for the crops is becoming ever more important, but also more difficult in view of the changing rainfall cycles.

Locally, two types of farmers are distinguished on the basis of their sowing cycles: *aggetas* (early sowers) and *pacchetas* (late sowers). Being a *paccheta* used to be a strategic choice for farmers in the past, depending upon farmers' decisions to keep their farms fallow or ploughing the stubble, or harvest late. However, *paccheta* now has become a synonym for laziness in farming. For the *rabi* wheat crop, a *paccheta* would harvest the wheat in March or April. Farmers have identified this period as especially risky and prone to rainfall, and responded to this risk by becoming all *aggetas*, hastily harvesting the rice crop and sowing the wheat with very little time in between. Farmers often resorted to stubble burning in their hurry to clear the field. However, after this practice had been declared illegal in light of the air pollution in nearby Delhi —a prohibition that is strictly enforced in Budhera— stubble burning has become rare.

The winter season of 2015 was marked by prolonged heat and a late onset of cold weather, as well as unseasonal rains in December. As a consequence, mustard, which is the second important winter crop of Budhera, suffered a double blow. *Aggeta* mustard farmers sowed their fields in October, hoping for low temperatures by November. However, the extended warm period during the first half of November led to early flowering in mustard plants, which impeded further growth and lowered the yields. Later, in December, the unseasonal rain caused the fields to be wet when the *paccheta* mustard was to be sown, and farmers had to wait for the soil to dry. Once they realized they would be late, most of them decided to leave the fields fallow or plant *jowar* (fox-tail millet) instead. Here both the *aggeta* and *paccheta* farmers faced different problems in sowing mustard.

The late onset of winter and the extended periods of sunshine were beneficial to some farmers. Those whose fields are located in low-lying areas near the GWS and NCR canals, which have very high-water tables from seepage, are generally the *pacchetas,* as the soil in their fields remains moist and takes time to dry. The moist soil also affects the harvesting of rice. In the harvesting process, rice plants are tied into stacks and kept on the field for 2–3 days before they are taken to the mill. With a wet soil, farmers aren't able yet to harvest the rice as it would spoil on the moist ground. However, with the prolonged sunshine, the farms near the canals had time to dry out. The farmers sowed wheat at the same time as farmer on other fields and were not the *pacchetas* for that year. The impact of climate variability thus plays out differentially at a micro-level, depending on various contingent factors.

The cumulative impact of these factors of urbanization (damage from the canals and wastewater irrigation) and climate variability or change is negotiated by farmers; not only in short-run decisions about cropping calendars but also impacting more stable long-run configurations like the fixing of rents for sharecropping. As mentioned, it is common for landowners in Budhera to rent their land to sharecroppers under an informal tenure system, locally known as *Kann-bataai*. For the after-harvest payment system, the rates paid by tenants have gone down as a consequence of reduced productivity and greater risks in agriculture. For a long time, the rent in *kann-bataai* used to be stable around 40 *man*[5] of wheat per acre for a year. From around 2014, the rates have generally gone down to 25 *man*.

There is also a structural shift in the institution of *kann-bataai*, or how sharecropping functions in the village. *Kann-bataai* labourers seldom work small plots of land through a sharecropping arrangement nowadays. This has become an unfeasible arrangement in Budhera, and has practically vanished. The system is now mostly restricted to large sharecroppers who manage teams of labourers working large plots of land rented from either a single large landowner or from many smaller landowners.

The greater scale of sharecropping operations may help manage risks in mild shocks to production but is of no use in major climate shocks. One of the farmers who had lost much from the unseasonal rains of March 2015 was a landowner-cum-sharecropper. He had three acres of land of his own, and 11 acres from another farmer through *kann-bataai*, with an upfront payment of INR 150,000 (or $2172). His harvest from a combined cultivation of 14 acres was around 60 quintals, which was an abysmal return for him. In case of crop failure, the loss is borne by the tenant and not the landowner; state policy fails to recognize informal tenants and only compensates landowners.

The peri-urban nature of Budhera rubs off on its agriculture in multiple ways. An interesting, seemingly trivial, detail about agriculture in Budhera was that most large farmers (unlike farmers in other districts of Haryana) don't own their tractors. The price of renting a tractor in Budhera is much lower than in other districts as a collateral effect of the heavy construction activities in Gurugram. At a rental of Rp. 300 per hour, during which you can get four acres ploughed, farmers opt for renting. Many villagers in Budhera and nearby peri-urban villages have become tractor entrepreneurs, who rent out tractors for agriculture for only a small percentage of the time, while mostly renting it out for construction, logistics and other non-farm activities around Gurugram.

6.6 Conclusion

In this chapter we have analysed the loss of common property resources and the emergence of urban canals in the peri-urban space as characteristics of the urbanization processes that are intensely operative in peri-urban Gurugram. We sought to

---

[5]A *man* (pronounced as rhyming with ton) is a local measure equivalent to 40 kg.

bring these processes in relation to changing agrarian conditions and climate change. Agrarian crisis and conflicts related to urbanization (manifest in land acquisition struggles) have been contentious political issues in India, and they feature much more prominently in political conversations than climate change. Meanwhile development programs and research agendas often use adaptation and resilience to frame policy narratives, including in studies related to the peri-urban, where the focus may deviate from ongoing conflict and struggle over water and land in the peri-urban. The focus on climate change often even tends to depoliticize these issues, as they become all framed in terms of adaptation rather than exploitation, inequality and power (Paprocki, 2018; Scoville-Simonds et al., 2020).

Although popular debates over access to land and water are slowly beginning to develop an explicit climate change formulation, they still do not adequately reflect the urgency of the climate crisis in India. The new realities of farming, such as the decreasing viability of small-holder *kann-bataai* arrangements and the breakdown of irrigation calendars, give rise to new forms of collective action by farmers. These measures of climate adaptation, often local in scale, are socio-ecological processes that are changing social relations among farmers and reflect even in political configurations of farmers' demands on the state. In the ongoing farmers' movement and agrarian politics of India, for instance, it has been observed that, in contrast to the powerful regional farmers' movement in India of the 1970s and 1980s, there has not been a split between the politics of large landowners on one hand and marginal farmers and landless labourers on the other (Jodhka, 2018). Climate change and emerging forms of adaptation among farmers are likely the undercurrents that have an important bearing on political alliances and discourses. Our research has elaborated on these adaptation processes, and future research will hopefully establish links between the political organization of farmers and the articulation of the agrarian crisis with adaptation as a political process.

Due to urbanization and climate change, agriculture is becoming uncertain and peri-urban residents are shifting to alternative livelihood opportunities in the nearby city. These changes are breaking the social fabric of the peri-urban, as not all residents (aged, women, many farmers) have the willingness to sell their land and take up new occupations. In an anticipatory urbanization approach, peri-urban farmers could be encouraged and supported by local authorities in developing sustainable forms of agricultural intensification. Such an approach could, for instance, create opportunities in the cities, using existing agriculture knowledge in innovative organic vegetable and food markets. City authorities and urban planners can also implement zoning regulations to ensure that farming areas, green spaces and water resources are protected from urban and non-agricultural rural development projects. Such policy actors need to understand that these not only affect peri-urban residents but also the urban and rural population and environments. Especially in the longer term these peri-urban issues might affect food production systems, agricultural ecology and agricultural knowledge among peri-urban youths. Budhera represents these peri-urban challenges, and we hope that in the near future the use of anticipatory urbanization approaches can contribute useful perspectives on sustainable peri-urbanization.

Finally, peri-urban research is necessary to unpack the relations between actors in re-defining, re-shaping, using, and managing their resources in the context of climate variability. Peri-urban spaces require contextualized approaches that can help understand processes of degradation and depletion of peri-urban CPRs and other resources, and of the emergence of new resources, such as wastewater canals, and related practices of use and management. These approaches should also emphasise that challenges in peri-urban areas are not only produced by urbanization and climate change, but also due to power dynamics, social relations and biased policy priorities (Narain et al., 2019; Roth et al., 2019). Currently, such an approach towards peri-urban spaces is completely missing in climate change adaptation policies, whether focusing on urban or on rural contexts.

References

Allen, A. (2003). Environmental planning and management of the peri-urban interface: Perspectives on an emerging field. *Environment and Urbanization, 15*(1), 135–148.

Allen, A., Hofmann, P., Mukherjee, J., & Walnycki, A. (2017). Water trajectories through non-networked infrastructure: Insights from peri-urban Dar es Salaam, Cochabamba and Kolkata. *Urban Research & Practice, 10*(1), 22–42.

Ballabh, V., Balooni, K., & Dave, S. (2002). Why local resources management institutions decline: A comparative analysis of van (forest) panchayats and forest protection committees in India. *World Development, 30*(12), 2153–2167.

Bowyer-Bower, T. (2006). *The inevitable illusiveness of "sustainability" in the peri-urban interface: The case of Harare*. Routledge.

Census India. (2001). *Census data 2001*. Planning Commission. Government of India. http://www.censusindia.gov.in/2011-common/censusdataonline.html. Accessed 9 Dec 2019.

Census India. (2011). *Provisional population totals*. Paper 1 of 2011: Haryana. Planning Commission. Government of India. http://censusindia.gov.in/2011provresults/prov_data_products_haryana.html. Accessed 20 Mar 2019.

Charmaz, K. (2006). *Constructing grounded theory: A practical guide through qualitative analysis*. Sage.

Colatei, D., & Harriss-White, B. (2004). Social stratification and rural households. In B. Harriss-White & S. Janakrajan (Eds.), *Rural India facing the 21st century* (pp. 115–159). Anthem.

Coward, E. W., Jr., & Levine, G. (1987). Studies of farmer-managed irrigation systems: Ten years of cumulative knowledge and changing research priorities. In *Public intervention in farmer-managed irrigation systems* (pp. 1–31). International Irrigation Management Institute (IIMI) and Water and Energy Commission Secretariat, Government of Nepal, Sri Lanka.

Eakin, H., Lerner, A. M., & Murtinho, F. (2010). Adaptive capacity in evolving peri-urban spaces: Responses to flood risk in the Upper Lerma River Valley, Mexico. *Global Environmental Change, 20*(1), 14–22.

Furlong, K., & Kooy, M. (2017). Worlding water supply: Thinking beyond the network in Jakarta. *International Journal of Urban and Regional Research, 41*(6), 888–903.

Gupta, D. (2005). Whither the Indian village: Culture and agriculture in "rural" India. *Economic and Political Weekly, 40*(8), 251–258.

Jodha, N. S. (1985). Population growth and the decline of common property resources in Rajasthan, India. *Population and Development Review, 11*(2), 247–264.

Jodha, N. S. (1986). Common property resources and rural poor in dry regions of India. *Economic and Political Weekly, 21*(27), 1169–1181.

Jodhka, S. (2018). Rural change in times of "distress". *Economic and Political Weekly, 53*(26), 5–7.

Johnson, R., & Onwuegbuzie, A. (2004). Mixed methods research: A research paradigm whose time has come. *Educational Researcher, 33*(7), 14–26.

Kundu, A. (2009). Urbanisation and migration: An analysis of trends, patterns and policies in Asia. *Human Development Research Paper* 2009/16. UNDP. https://mpra.ub.uni-muenchen. de/19197/1/MPRA_paper_19197.pdf

Lwasa, S., Mugagga, F., Wahab, B., Simon, D., Connors, J., & Griffith, C. (2014). Urban and peri-urban agriculture and forestry: Transcending poverty alleviation to climate change mitigation and adaptation. *Urban Climate, 7*, 92–106.

Manikandan, M. S., & Sundaram, K. (2017). Managing common property resources (CPRs) for rural development: A development perspective. *Productivity, 58*(3), 340–351.

Marshall, F., Waldman, L., MacGregor, H., Mehta, L., & Randhawa, P. (2009). *On the edge of sustainability: Perspectives on peri-urban dynamics* (STEPS working paper 35). STEPS Centre.

Mishra, P., & Narain, V. (2018). Urban canals and peri-urban agrarian institutions. *Economic and Political Weekly, 53*(37), 51–58.

Murthy, R. V. (2013). Political economy of agrarian crisis and subsistence under neoliberalism in India. *The NEHU Journal XI, 1*, 19–33.

Narain, V. (2014). Whose land? Whose water? Water rights, equity and justice in a peri-urban context. *Local Environment, 19*(9), 974–989.

Narain, V., & Nischal, S. (2007). The peri-urban interface in Shahpur Khurd and Karnera. *Environment and Urbanization, 19*(1), 261–273.

Narain, V., & Prakash, A. (2016). *Water security in peri-urban South Asia: Adapting to climate change and urbanization.* Oxford University Press.

Narain, V., & Singh, A. K. (2017). Flowing against the current: The socio-technical mediation of water insecurity in peri-urban Gurgaon. *Geoforum, 81*, 66–75.

Narain, V., Ranjan, P., Singh, S., & Dewan, A. (2016). Urbanization, climate change and water security in peri-urban Gurgaon, India. In V. Narain & A. Prakash (Eds.), *Water security in periurban South Asia: Adapting to climate change and urbanization.* Oxford University Press.

Narain, V., Vij, S., & Dewan, A. (2019). Bonds, battles and social capital: Power and the mediation of water insecurity in peri-urban Gurgaon, India. *Water, 11*(8), 1607–1619.

Ong, A. (2006). *Neoliberalism as exception: Mutations in citizenship and sovereignty.* Duke University Press.

Ostrom, E. (1992). *Crafting institutions for self-governing irrigation systems.* Centre for Self Governance.

Paprocki, K. (2018). The anti-politics of climate change. *The Daily Star*, 8 June 2018. https://www. thedailystar.net/star-weekend/environment/anti-politics-climate-change-1588075

Piorr, A. (Ed.). (2011). *Peri-urbanisation in Europe: Towards European policies to sustain urban-rural futures; Synthesis report; PLUREL [sixth framework programme].* Forest & Landscape, University of Copenhagen.

Preston, B. L., Westaway, R. M., & Yuen, E. J. (2011). Climate adaptation planning in practice: An evaluation of adaptation plans from three developed nations. *Mitigation and Adaptation Strategies for Global Change, 16*(4), 407–438.

Rakodi, C. (1999). Poverty in the peri-urban Interface. In *Natural resources systems Programme (NRSP) research advances 5* (pp. 1–4). DFID.

Ranganathan, M., & Balazs, C. (2015). Water marginalization at the urban fringe: Environmental justice and urban political ecology across the North–South divide. *Urban Geography, 36*(3), 403–423.

Ranjan, P., & Narain, V. (2012). *Urbanization, climate change and water security: A study of vulnerability and adaptation in Sultanpur and Jhanjhrola Khera in peri-urban Gurgaon, India* (Peri-urban water security discussion paper series, paper no. 3). SaciWATERs, India.

Ravetz, J., Fertner, C., & Nielsen, T. S. (2013). The dynamics of peri-urbanization. In *Peri-urban futures: Scenarios and models for land use change in Europe* (pp. 13–44). Springer.

Roth, D., Khan, M. S. A., Jahan, I., Rahman, R., Narain, V., Singh, A. K., … Yakami, S. (2019). Climates of urbanization: Local experiences of water security, conflict and cooperation in peri-urban South-Asia. *Climate Policy, 19*(sup1), S78–S93.

Scoville-Simonds, M., Jamali, H., & Hufty, M. (2020). The hazards of mainstreaming: Climate change adaptation politics in three dimensions. *World Development, 125*, 1040683.

Selco Foundation. (2016). *The Indian urban migrant crisis- A need for inclusivity.* https://www.selcofoundation.org/the-indian-urban-migrant-crisis/

Sharan, A. (2014). *In the city, out of place: Nuisance, pollution, and dwelling in Delhi, C. 1850-2000.* Oxford University Press.

Sridhar, K. (2010). Impact of land-use regulations: Evidence from Indian cities. *Urban Studies, 47*, 1541–1569.

Swyngedouw, E. (1997). Power, nature, and the city: The conquest of water and the political ecology of urbanization in Guayaquil, Ecuador, 1880–1990. *Environment and Planning A, 29*(2), 311–332.

UNFCCC, Cancun Adaptation Framework. (2010). *Adaptation: The Cancun agreements.* United Nations Framework Convention on Climate Change. https://unfccc.int/tools/cancun/adaptation/index.html

Vij, S. (2014). Urbanization, common property resources and gender relations in a peri-urban context. *Vision, 18*(4), 339–347.

Vij, S. (2019). *Power interplay between actors in climate change adaptation policy-making in South Asia.* Ph.D. dissertation, Wageningen University & Research.

Vij, S., & Narain, V. (2016). Land, water and power: The demise of common property resources in periurban Gurgaon, India. *Land Use Policy, 50*, 59–66.

Vij, S., Narain, V., Karpouzoglou, T., & Mishra, P. (2018). From the core to the periphery: Conflicts and cooperation over land and water in periurban Gurgaon, India. *Land Use Policy, 76*, 382–390.

von Benda-Beckmann, F., von Benda-Beckmann, K., & Wiber, M. (2006). The properties of property. In F. von Benda-Beckmann, K. von Benda-Beckmann, & M. Wiber (Eds.), *Changing properties of property* (pp. 1–39). Berghahn Books.

White, J. M. (2016). Anticipatory logics of the smart city's global imaginary. *Urban Geography, 37*(4), 572–589.

Yadav, Y. (2017). *Interview: We are witnessing the beginning of a peasant rebellion in India, says Yogendra Yadav.* Scroll.in, 4 October, 2017. https://scroll.in/article/851846/interview-we-are-witnessing-the-beginning-of-a-peasant-rebellion-in-india-says-yogendra-yadav

Yarina, L. (2018). Your sea wall won't save you: Negotiating rhetorics and imaginaries of climate change. *Places Journal.* https://placesjournal.org/article/your-sea-wall-wont-save-you/?cn-reloaded=1

Chapter 7
Views from the Sluice Gate: Water Insecurity, Conflict and Cooperation in Peri-Urban Khulna, Bangladesh

M. Shah Alam Khan, Rezaur Rahman, Nusrat Jahan Tarin,
Sheikh Nazmul Huda, and A. T. M. Zakir Hossain

7.1 Introduction

Management of freshwater, wastewater, storm runoff and river salinity has played a central role in the emergence and evolution of conflict and cooperation in the urban and peri-urban areas of Khulna. This water management largely depends on how and to what purpose the water infrastructure, in this case more specifically sluice gates, have been designed and are operated and maintained, and to what extent these meet the needs and priorities of communities and other stakeholders. Conflicts as well as new alliances and forms of cooperation arise in specific ways, where sluice gate operation is influenced or controlled by powerful groups to support their own interests. Conflicts and cooperation take on different forms and are manifested in stakeholder relations, while their characteristics vary in temporal and spatial scales. Water insecurity of the peri-urban population is closely related to these conflicts and cooperation, as seen across South Asia (Narain et al., 2013; Roth et al., 2018). These water insecurities and conflicts are aggravating because of the compounded effects of urbanization and climate change (Khan et al., 2016).

In rural Bangladesh sluice gates often give rise to conflicts, mainly between farmers and fishers, about the timing of operation and amount of water passing through the sluice (Murshed & Khan, 2009). In this chapter we analyse the dynamics of conflict and cooperation in peri-urban Khulna around a sluice gate at the outfall of the Mayur River passing through Khulna. The sluice gate and the adjoining embankments were constructed primarily for stormwater management, storm surge protection, wastewater discharge and river salinity control for the upstream

M. S. A. Khan (✉) · R. Rahman · N. J. Tarin
Institute of Water and Flood Management, Bangladesh University of Engineering and Technology, Dhaka, Bangladesh

S. N. Huda · A. T. M. Z. Hossain
Jagrata Juba Shangha, Khulna, Bangladesh

V. Narain, D. Roth (eds.), *Water Security, Conflict and Cooperation in Peri-Urban South Asia*, https://doi.org/10.1007/978-3-030-79035-6_7

123

urban areas. The gate was later modified to provide additional agricultural benefits to the downstream peri-urban areas, where the Mayur River was important for agriculture, fish farming and open-water fisheries. Therefore proper operation of the sluice gate has always been vital for peri-urban livelihood sustenance. Peri-urban livelihoods are not only threatened by the absence of an equitable and optimal operation of the sluice gate to meet the urban and peri-urban requirements, but also by urban expansion, loss of water bodies and upstream wastewater discharge. Being converted into housing projects, peri-urban agricultural lands make room for urban expansion. In addition, overall environmental degradation is observed as a consequence of solid waste dumping and wastewater discharge into the river without adequate treatment to remove harmful materials and chemicals (Dasgupta et al., 2016; Khan et al., 2016).

The sluice gate, primarily serving the city but located in peri-urban Alutala (Tentultala Village), created conflicts between urban and peri-urban water users. Conflicts started to emerge around the sluice gate as its control was taken over covertly by powerful urban elites[1] together with local powerful groups, to ensure saline water availability in their shrimp farms. Such gate operation deprived the peri-urban population of freshwater needed for subsistence agriculture and freshwater fisheries. Further, rapid and unplanned urbanization at the cost of lowlands and water bodies also started to pose threats to the natural flow of the Mayur River. Untreated solid wastes and wastewater from domestic, industrial, commercial and clinical sources in Khulna city accumulated in the river, which eventually exceeded its assimilation capacity. Urban wastewater coupled with increased salinity, especially in the dry season, makes the water in the peri-urban area unusable for irrigation and aquaculture. Maintenance of adequate flow in the river through proper gate operation could ensure flushing of the wastewater, but such a gate operation rule was never institutionalized.

In this chapter we explore the different forms of conflicts and cooperation, how and why they have changed over time, and how the communities have mobilized to deal with these conflicts. In our analysis we focus on the sluice gate, its function and control, and implications for the peri-urban communities. We view the hydrological, social and institutional processes as an integrated hydro-social continuum, in which these apparently separate processes interact with and influence one another, and are closely interlinked and intertwined. On the one hand, hydrological requirements and water disasters shape people's lives, livelihoods and social relations. On the other hand, the hydrological system and its alterations are influenced by various socio-economic priorities, institutions and power relations. Similar interdisciplinary approaches to the interlinked social and hydrological processes in relation to negotiations, dialogues and conflict mitigation can be found in the literature (Boelens et al., 2016; Massuel et al., 2018, 2019; Wesselink et al., 2017).

---

[1] People living in the city and having decisive influence on urban policies and programs through their political connection. We more specifically refer to the people with business interests in the urban and peri-urban areas, particularly fish farming in peri-urban areas.

In the rest of the chapter we first describe the approach and methods followed in the study, including an explanation of our hydro-social conceptualization of the setting, methods of research and interventions, rationale for the selection of the study site, and contextualization of our action research approach. This is followed by a description of the peri-urban water insecurity problems and of the critical issues specifically related to the sluice gate, including hydrological and historical contexts, major actors and contestations around the gate. We then present a detailed analysis of major conflict cases, to explore the roles and power of the main actors in these cases centering on the sluice gate, the contradictory interests of the stakeholders, and their varying views of problems and solutions. In the subsequent section we present an overview of our action research to address these issues, reflecting on our experience in dealing with power relations and interest networks.

7.2 Research Approach and Method

Water insecurity in peri-urban Khulna is primarily a consequence of the urban policy that prioritizes urban services and amenities over peri-urban requirements (Khan et al., 2016). The sluice gate at Alutala is an example of such policy priority, in which peri-urban needs are ignored if not absent. The gate, located in the peri-urban area, was constructed to protect the city from storm surges and salinity intrusion, and to flush out urban wastewater and storm runoff. Peri-urban agriculture and freshwater availability were of secondary importance in the original design considerations for the gate. The gate was modified to meet these peri-urban hydrological requirements at a later stage.

Thus, in the analysis of water insecurity, conflict and cooperation in peri-urban Khulna, the role of urban institutions and actors that influence the water flow and discharge in the peri-urban area is important. After the construction of the gate, its control became attractive, as the urban and peri-urban power alliances saw benefits in altering the peri-urban hydrology and water bodies. We therefore view the sluice gate as central to a hydro-social continuum in the analysis of conflict and cooperation in relation to water insecurity, in which water, people, institutions and actors are interlinked and interdependent. Contestation over water and land evolves through hydro-social processes, as the peri-urban space becomes increasingly lucrative in terms of economic returns from these alterations of the hydrological system.

In the research methodology we adopted a predominantly qualitative approach. Semi-structured interviews and key informant interviews were conducted to explore the main conflict cases and to identify key actors and stakeholders. Subsequent focus group discussions with urban and peri-urban communities were carried out to further analyse and understand the historical emergence and evolution of conflicts around the sluice gate. Stakeholder analysis and social power mapping were conducted at two stages to analyse the stakeholder arrangements and power relations. In the first stage, this was conducted at the research sites with smaller focus groups. In the second stage, the stakeholder diagrams and social power maps prepared in the first stage were triangulated at a workshop with a broader stakeholder group. The research team made several field visits for direct observation of the hydrological setting, water management infrastructure and biophysical issues raised during the interviews and discussions. The research team also collected water samples from the river for laboratory analysis. These direct observations and water quality analyses were verified with secondary hydrological data and catchment information. Outcomes of these analyses and explorations were utilized to create avenues and platforms to improve the capacity of the marginalized communities to negotiate and overcome their challenges to reach more cooperative solutions. This was attempted through a series of community meetings and stakeholder workshops.[2]

7.3 Research Location

Khulna city, located in the south-western hydrological region of Bangladesh, is home to around 1.4 million people (in 2011) (Bangladesh Bureau of Statistics, 2011). Socio-economic developments in the south-west coastal region are leading to new industries and human settlements in Khulna city and its surrounding areas, and changing the land use pattern and local resources. Export processing industries based on shrimp farming, in particular, are strengthening the economy in Khulna and its surroundings (Khan et al., 2016). Khulna is frequently confronted by severe water problems and disasters, including cyclones, storm surges, floods, high-intensity rainfall, water logging, saline intrusion and river erosion. The increasing trends in the frequency and severity of cyclones, storm surges and high-intensity rainfall events, along with increasing river salinity because of sea level rise, are attributed to global climate change (Khan et al., 2013). Notably, Khulna is among the fifteen most vulnerable cities under the impacts of climate change (Hallegatte et al., 2013; Hanson et al., 2011). The devastating cyclones *Sidr* in 2007 and *Aila* in 2009 have forced people to migrate to the peri-urban areas of Khulna. This in-migration has created additional competing demand for freshwater from already-stressed sources, and has added to wastewater generation. Climate change is also believed to have increased high-intensity rainfall events in the Khulna area (Mondal et al., 2013). These short rainfall spells create spikes of urban floods and release high volume of rainfall-runoff that has to be drained through the Mayur river and eventually through the sluice gate.

---

[2] The research presented here was executed in the framework of the NWO/DFID-funded research programme "Conflict and Cooperation in the Management of Climate Change (CoCooN/CCMCC)". The research team consisted of members from Bangladesh University of Engineering and Technology (BUET) and Jagrata Juba Shangha (JJS), a local NGO. BUET carried out the scientific studies and analysis, while JJS was responsible for community mobilization and action components of the research. This research built on a previous action research project funded by the Canadian International Development Research Centre (IDRC), which explored water security in peri-urban Khulna (Khan et al., 2016). The team was also closely linked with another research project dealing with groundwater security in peri-urban Khulna, which was funded by the Urbanising Deltas of the World program of the Dutch Research Council (NWO).

The Mayur River, which flows through Khulna city and hydrologically connects the urban and peri-urban areas, originates from the natural wetlands *Beel Pabla* and *Beel Dakatia*, and drains into the Rupsha-Bhairab River near Alutala (a peri-urban location) (Fig. 7.1). The Mayur River receives water from a drainage area of about 53 square kilometers, including the Khulna City Corporation (KCC) area. The river serves as an urban amenity, and provides ecosystem services, water for domestic, industrial and commercial uses, and irrigation water for peri-urban farmers. It also provides other benefits, such as recreation and tourism functions, climate and flood regulation, biodiversity sustenance, nutrient re-cycling and waste assimilation.

The Bangladesh Water Development Board (BWDB) constructed the Khulna city protection embankment in the late 1970s. As a part of the embankment system, a sluice gate was constructed at the outfall of the Mayur River near Alutala in the early 1980s (Fig. 7.2). Ten flap gates were included in the original design of the sluice structure to protect Khulna city from storm water flooding, storm surge and salinity intrusion. Flap gates open automatically when the internal water level is higher and close when the external water level is higher. The sluice gate is locally known as "*Dosh* (ten in Bangla) gate".

As peri-urban agriculture and fish farming became more important, BWDB added ten vertical gates to the original structure for the control and retention of internal water storage. This modification of the structure facilitated peri-urban agriculture with greater freshwater availability in the dry season. The saline and freshwater fish farms in ponds and an abandoned channel of the Mayur River also started to gain interest. Figure 7.3 shows the physiography and landmarks around the sluice gate. As the fish farm owners grew in number and power, they started

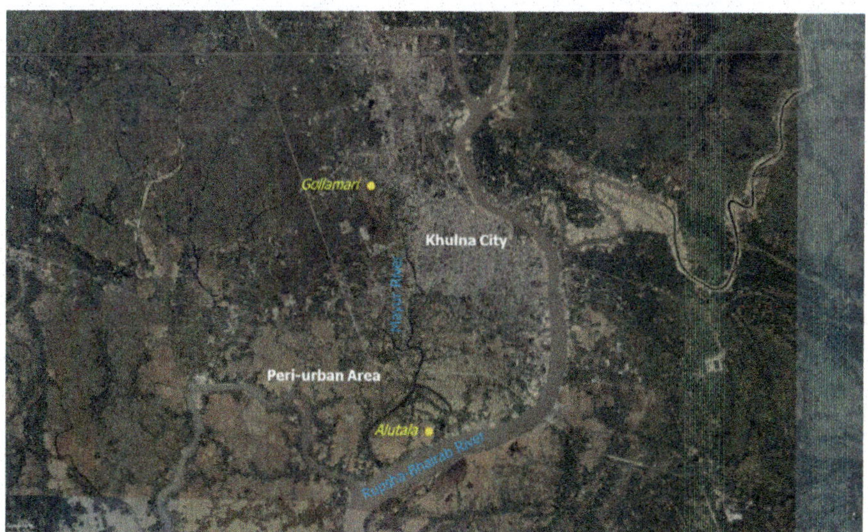

Fig. 7.1 Google map showing the Mayur River, Khulna city and the peri-urban area

Fig. 7.2 The sluice gate at Alutala. (Photo Jagrata Juba Shangha)

Fig. 7.3 Google map showing the sluice gate on the main channel (Kazi Bacha River) and fish farms on the abandoned channel of the Mayur River

to control the vertical gate operation to store saline water for their fish farms. This gate control reduced the capacity of the gate to discharge urban wastewater flowing from the upstream urban areas. Thus the operation and management of the sluice gate had both socio-technical and agro-environmental implications for the water use conflicts, many of which emerged after the sluice gate had been designed and constructed without considering the peri-urban water requirements.

The Mayur River was a free-flowing tidal river some 40 years back. Trawlers and large country boats used to ply the river with passengers and goods. Since then, disposal of solid wastes and discharge of wastewater from the urban areas have severely degraded the conveyance and assimilation capacities of the river. The discharge of untreated wastewater through some 22 urban outfalls and the dumping of the solid wastes from the nearby slaughterhouses, markets, bus terminal, hospitals, clinics, automobile factories and industries have filled up the river and increased its pollutant loads. A large slaughterhouse located on the river bank discharges wastewater directly into the river without any treatment. Encroachment of the river has reduced its width, turning it into a narrow wastewater channel at many places such as Rayermahal, Gollamari and Shashanghat. At Gollamari point, the river depth is only 0.3 m (see e.g. Khan et al., 2016).

Because of its reduced retention, conveyance and discharge capacities, the Mayur River has become too polluted to support aquatic lives and peri-urban livelihoods. Water quality analysis indicates that dissolved oxygen, biochemical oxygen demand, total dissolved solids (TDS) and salinity levels along most of the river stretch do not meet the recommended limits for drinking and irrigation water quality set by the Department of Environment of Bangladesh and the World Health Organization. During the dry season, the water level of the river drops drastically and its pollution level rises. Consequently, the fisheries and other aquatic resources in the river deplete, resulting in loss of livelihoods of the peri-urban people. Growing land development businesses in Khulna have filled up open water bodies and peri-urban lowlands. As urbanization has increased the domestic and industrial water demands and depleted the surface and groundwater resources, the competition for these scarce water resources among different user groups in the urban and peri-urban areas has also increased. This competition has created complex water use conflicts between the urban and peri-urban communities.

Drinking water supply in Khulna and its surrounding areas mainly depends on groundwater replenished from the hydraulically connected rivers. Thus quantity and quality of the river water and withdrawal rate from the groundwater aquifers have direct implications for groundwater supply. Secondary data indicate that the groundwater contains a relatively high level of salinity, which often makes the water unpleasant or unsuitable for drinking and domestic uses. Observed salinity levels in the groundwater represented by TDS indicate that groundwater salinity is generally higher during the dry season. Excessive groundwater withdrawal is partly blamed for increased salinity intrusion into the local aquifers. The hand tube wells (HTW) installed to withdraw water from the shallow (upper) aquifer run dry during the summer (March–May) months. Presence of iron and arsenic in excess of the drinking water standard in the upper aquifer groundwater is another constraint to its use. The Khulna Water Supply and Sewerage Authority (KWASA) can supply piped water to only 30% of the urban population. The rest of the urban and peri-urban population depends on shallow HTWs, deep tube wells (DTW) and surface water bodies such as ponds, canals and rivers. The unplanned and rapid urbanization of the city has resulted in a decreasing area of lowlands, ponds, lakes and other water bodies, as well as a reduction in agricultural land. Urbanization has thus reduced the groundwater recharge areas and increased the peak of storm runoff. Consequently, waterlogging created in many places of the city and its peri-urban areas in the event of high-intensity rainfall causes immense suffering to the people. Inadequate drainage facilities and blockage of drains due to improper solid waste management and maintenance further aggravate the drainage of storm runoff.

The urban storm water and wastewater management priorities together with pressure from the urban environmental activist groups forced KCC to negotiate with BWDB to ensure an optimal operation of the sluice gate that would also meet the peri-urban requirements. At the same time, the peri-urban communities mobilized to protest the control of the sluice gate by fish farm owners. The responsibility of sluice gate operation was eventually handed over to KCC in 2012. KCC formed a Gate Operation Committee (GOC), which included members from KCC, Khulna Development Authority (KDA), BWDB, community representatives and government agencies. The GOC was mandated to decide the gate operation time and duration, in consultation with the downstream communities. In reality, however, this decision was taken jointly by a KCC official who owns agricultural land in Alutala and the local *Union Parishad* chairman.[3] Despite this new arrangement for gate operation, apparently in favor of the peri-urban communities, the water quality of the Mayur River did not improve significantly. BWDB, still responsible for repair and maintenance of the gate, keeps deferring funds required for the necessary repairs of the structure. As a result, the sluice gate is not operational as required. Meanwhile, there has been an increased interest in cooperative fish farming among the peri-urban communities, with the blessing of local political leaders. Fish farm owners in the abandoned channel are moving to other parts of the river where the water quality is better, while peri-urban farmers are adapting to less water-demanding crops.

7.4 Conflicts Around the Sluice Gate

7.4.1 Function of the Sluice Gate

Livelihoods and ecology in peri-urban Khulna have been largely dependent on the connectivity between the Rupsha-Bhairab and Mayur rivers. Construction of the embankment and the sluice gate disrupted local livelihoods and agricultural practices. Flow of the Mayur River not only links the peri-urban area with the upstream urban area hydrologically but also relates these two spaces, separated by a policy divide, through several shared issues and concerns. Khulna city residents enjoy the priority in formal policy making and implementation through KCC and other public agencies, while the peri-urban communities, in the absence of formal policy recognition,

---

[3] Smallest rural administrative and local government unit in Bangladesh. A Union consists of several villages.

struggle to have their needs attended to. Regulation of the Mayur River flow through the sluice gate has direct and indirect implications for these shared issues and concerns.

While the sluice gate primarily serves the urban area for wastewater and storm water management, the downstream peri-urban concerns of water quality, water availability and water logging are ignored in gate operation. Degrading river water quality has forced the peri-urban communities to rely more on groundwater for drinking and irrigation. The gate operation, often influenced by powerful groups, marginally favors local forms of collective action such as co-operative fish farms, but only at the mercy of local political leaders. The main areas of conflict around the sluice gate are summarized below.

7.4.2 Urban Versus Peri-Urban Needs

The urban and peri-urban areas linked by the Mayur river have distinctly different water use priorities. Mitigation of wastewater pollution and storm water flooding are the important concerns for the city, whereas availability of freshwater for irrigation, domestic use and fisheries is important for the peri-urban area. The sluice gate was designed and constructed to serve urban priorities. The GOCs formed by both BWDB and KCC operated the gate in favor of the urban needs. Khulna city is situated on higher ground upstream of the river. Therefore, to maintain flow and usable water level in the urban segment of the river, the gate has to be operated in such a way that water is not discharged through the gate too fast. This results in a relatively high water level in the downstream peri-urban areas, often causing flooding and water logging in areas near the gate. The wastewater generated in the city is flushed to the Rupsha river through the sluice gate. In order to dilute the internal polluted water, saline water is brought in during high tides by forcing the external flaps open, which is against the design operation mode. While the saline water fish farms, mostly owned by urban businessmen, utilized and manipulated this opportunity, the peri-urban communities were largely deprived of their freshwater needs. Saline water is also brought in to kill the internal water hyacinth growth. The "gate man", appointed by BWDB and later by KCC, holds the key for gate operation as per the instruction of the GOC. However, the actual operation of the gate by the gate man is often influenced by the fish farm owners.

7.4.3 Open-Water Capture Fisheries Versus Fish Farming

Prior to the construction of the sluice gate, open-water capture fisheries was the main source of livelihood for the peri-urban communities. Only 20–30% of the population owned agricultural land and had crop farming as their primary means of livelihood. Construction of the city protection embankment and the sluice gate disconnected the floodplains of the Mayur river and its tributaries from the Rupsha river. This obstruction severely affected the fisheries resources in the internal water bodies. Since open-water fishing in the much larger Rupsha river needed higher investments, most open-water fishers were forced to start small-scale aquaculture in leased ponds and lowlands adjacent to the river. Large-scale fish and shrimp farming started in the early 1990s in the leased *khas* lands[4] and illegally grabbed lowlands. Most of these farms were financed by politically empowered urban businessmen and run by powerful local groups. This alliance of power and money dominated fish farming in the area and started to control operation of the sluice gate in its favor. This dominance also forced many people to either give up small-scale fish farming or work as day laborers in a large farm.

7.4.4 Peri-Urban Agriculture Versus Fish Farming

Large-scale saline water fish farming also affected the livelihoods of the small farmers. Commercial fish farms started to spread in the abandoned channel of the Mayur river, encroached canals and lowlands. Although some of the encroachments and illegal grabbing of *khas* lands were evicted occasionally, re-encroachment took place as soon as there was a change in the position of the government official implementing the eviction order. The large fish farms influenced the operation of the sluice gate to bring in saline water to their farms. Such control of the gate deprived farmers of the irrigation water they needed. At the same time, polluted discharge from the fish farms and increasing soil salinity around the fish farms posed threats to crop cultivation. This situation changed later: as the peri-urban community mobilized to protest saline water fish farming, the commercial fish farms started to lose their interest because of degrading water quality, and a community representative was included in the GOC.

To explore the role and involvement of the stakeholders around sluice gate operation, we conducted a stakeholder analysis in two stages. A social power matrix was prepared to understand the power relations between these stakeholders. First, a list of key stakeholders (given below) was prepared through key informant interviews and focus group discussions.

1. Peri-urban community, fishers, farmers
2. Peri-urban fish farms
3. Gate Operation Committee (GOC)
4. Khulna City residents
5. Khulna City Corporation (KCC)
6. Bangladesh Water Development Board (BWDB)
7. Local Government Institutions (LGI)
8. Khulna Development Authority (KDA)
9. Khulna Water and Sewerage Authority (KWASA)
10. Department of Environment (DoE)

---

[4]Land owned by the government, and available for allocation according to government policy.

11. Department of Agricultural Extension (DAE)
12. Community Based Organizations (CBOs)
13. Non-Government Organizations (NGOs)
14. Water and Goods transport businesses
15. District Administration (DA)
16. Khulna University (KU)
17. Khulna University of Engineering and Technology (KUET)
18. Bangladesh University of Engineering and Technology (BUET)
19. Media (local and national)

Subsequently, a series of smaller group discussions at the local sites and a larger meeting in Khulna were conducted to collectively understand the key stakeholders' involvement. In these discussions, the key stakeholders were identified in four groups according to their primary role in gate operation and management: users, decision makers, implementers and experts, and were placed in three tiers: co-operating, co-thinking and co-knowing, according to their degree of involvement. Co-operating stakeholders are the most closely involved in the issue, through active interaction with the other stakeholders. Co-thinking stakeholders take part in passive consultation with other stakeholders but do not interact actively with others. Co-knowing stakeholders are only aware of the issue and do not actively participate in the interactions. A stakeholder diagram (Fig. 7.4) shows this distribution.

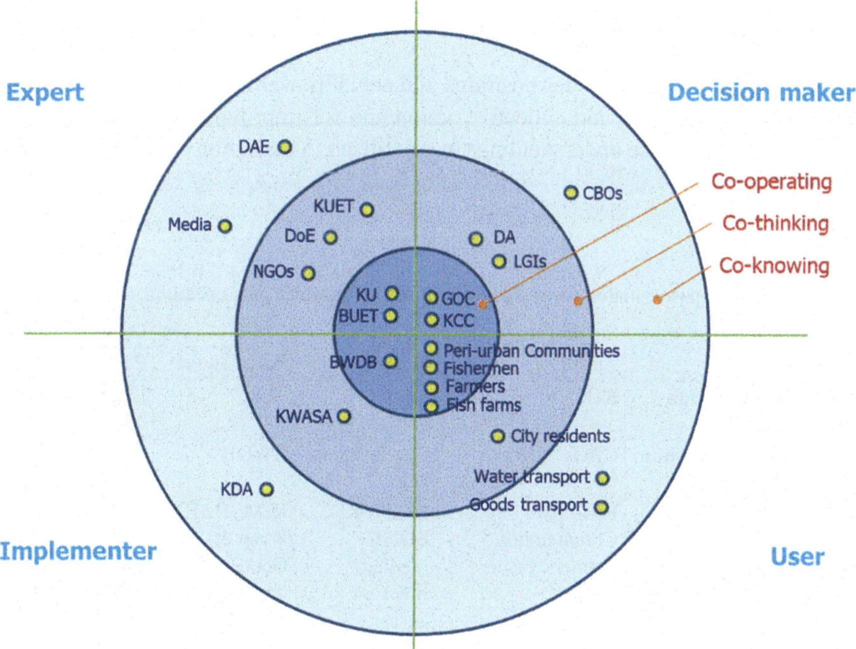

Fig. 7.4 Stakeholder alignment around sluice gate operation and management

On the basis of these, the social, political, institutional and bureaucratic power of the stakeholders to influence the operation of the sluice gate was further explored. The social power matrix (Table 7.1) shows the arrangement of the stakeholders according to their position with regard to an equitable and optimal operation of the sluice gate, and their level of social power to influence any change.

The stakeholder diagram and the social power matrix together represent the manifestation of the conflicts around the sluice gate in stakeholder positions and power relations. Despite KCC having the most active role and interest in the issue, its organizational capacities are largely limited because of a passive role of BWDB, which still owns the physical structure and is responsible for its repair and maintenance. Although the GOC has been mandated on the basis of an apparent consensus to operate the gate, its actual actions and implementation of decisions are influenced behind the scene by the LGIs, who also have stakes in the fish farms. The DA often plays a strong role in evicting the canal grabbers, but does not have a significant role in gate operation. Public agencies like KWASA and KDA are involved with drainage of greater Khulna, but not with sluice gate operation. DoE monitors the salinity and water quality parameters for river conservation, and can assist in establishing a gate operation rule. DAE works at the Union level to advise on eco-friendly, climate-resilient and sustainable agricultural practices. It can assist farmers by suggesting alternative suitable agricultural practices according to the available quantity and quality of water. The peri-urban communities, more specifically farmers and fishers, are most severely affected because of the construction and inequitable operation of the gate. They also struggle unsuccessfully in the margins of the social power structure.

The analysis of stakeholder positions and social power also revealed the opportunities for cooperative and collective action and for improving the gaps in stakeholder relations. These understandings were utilized in the action components of the research.

Table 7.1 Social power matrix for stakeholders influencing sluice gate operation

		Support			Opposition	
		Active	Passive	Fence-sitters	Passive	Active
Stakeholder Power	**High**	KCC	DA	GOC	LGIs	Fish farms
	Medium	CBOs, NGOs	City residents DoE	BWDB		
	Low	Peri-urban Communities, Farmers, Fishers	KUET, KU, BUET, Media, KWASA	KDA, DAE, Water & Goods transport		

7.5 Dealing with Sluice-Related Conflicts

7.5.1 Sluice Gate Operation

The research team has been actively involved at various stages in advocating an equitable operation of the sluice gate and institutionalizing a gate operation rule acceptable to all stakeholders, starting from the previous IDRC-funded research. A particular focus of the research was on mitigating conflicts around the sluice gate through capacity building of the marginalized peri-urban communities. From the very initial period of sluice gate construction and operation, different stakeholders derived different benefits from this water infrastructure. The varying demands on its operation resulted in conflicts among the stakeholders. From time to time, measures were taken to reconcile parties in these conflicts. For example, the gate was erected to flush out urban stormwater and prevent salinity intrusion with a design of one-way water drainage, which created irrigation water scarcity in the dry season. Later, considering the water need for agriculture, BWDB constructed vertical gates that enabled the sluice gate to allow and retain water inside the gate by averting one-way drainage. Another area of cooperation made by BWDB was handing over the operation and management of the sluice to KCC. As KCC provides all public services including water supply, drainage, and wastewater management within the city, the structure was handed over to them to minimize bureaucratic procedures and to accelerate and upgrade public services for city people. The research team in the previous IDRC-funded project was instrumental in facilitating this transition. Although the peri-urban community was not a central consideration, this cooperation led to the selection of a well-off farmer of Alutala, who is also an official of KCC, as the secretary of the GOC. The village leaders expected that he could play a role in creating cooperation between local farmers and fishermen.

During our long-term engagement with the issue we see indications of an improved level of satisfaction among the peri-urban residents regarding the operation of the sluice gate. This improvement is partly because of the changes in gate operation favoring agriculture (including vertical gate to allow water into the river) and partly because of the shift in agriculture to less water-demanding crops. The combined problem of salinity and urban wastewater has resulted in a shift in cropping choices away from rice to less water-demanding winter crops. This is evident from farmers cultivating sesame, melon, bitter gourd etc., which require less water. Farmers cultivate *aman* rice once in a year during the rainy season. Informal interviews with farmers and residents suggest that this shift in agricultural practices has reduced the number of conflicts in water demand. Noticeably, shrimp farming by urban elites in the peri-urban areas has also decreased, due to protests by local farmers and fishermen. However, the cost of transformation in peri-urban agricultural practices as an option for conflict resolution needs to be analysed from an economic viewpoint as well. It might be economically viable to replace rice cultivation with winter vegetables, considering the demand for vegetables in the city.

The research team facilitated bringing all parties together on a common platform and formulating a gate operation rule to optimize the water needs of all stakeholders. After understanding the water requirements of different stakeholders and their power relations, the research team met with them separately to discuss the gate operation issue. Jagrata Juba Shangha (JJS), a local organization with a good network, was successful in building rapport and establishing open communication. In the next step, the stakeholders were brought together for a dialogue on the issue. It was, however, challenging to engage with the fish farm owners in the process for two reasons. First, they were reluctant to communicate with the local communities because of protests against the fish farms and a subsequent law suit. Second, the fish farms were already relocating to other parts of the river because of degrading water quality. Thus, although the fish farm owners participated in the dialogue, they did not have any strong opinion. Rather, members of the LGIs and local political leaders, who previously had stakes in the fish farms and manipulated gate operation, saw new opportunities in community co-operative fish farms, and supported gate operation in their favor. As the gate remains relevant mostly for wastewater discharge and flood regulation, all parties eventually agreed and decided through interactive discussion that the gate should be operated in such a way that the water level remains 2 ft below the bank of the river in the downstream peri-urban area. This would be an acceptable solution for farmers and fishermen, and for groundwater recharge as well.

Our most recent investigations (in 2020) found that the GOC has become less active, since the KCC official and a peri-urban farmer representing KCC in the GOC have been replaced by another person who is higher in authority. Currently the gateman directly follows his orders for gate operation without consultation in the GOC. The new KCC official in the GOC, having a higher authority and not having a personal stake in the peri-urban area, is less interested in community consultation. The other GOC members also feel less comfortable to communicate with a person of higher authority. At the same time, the local communities are losing interest in the negotiation platform and dialogue, as the water quality of the river has become so poor that it has become unusable for any purposes. Hence the conflicts around gate operation are becoming moot.

7.5.2 Capacity Development

The objectives of the capacity development activities were to build the capacity of the stakeholders to understand peri-urban water security concerns and use the research outcomes and learning to sensitize state actors, professionals and communities. JJS, the local partner in the research team, played an active role in this regard. The capacity of JJS itself was first developed through its participation in various meetings, workshops and training programs of the project. These included capacities to analyse and mediate conflicts, and a more holistic understanding of the socio-technical complexities. Then JJS became instrumental in developing the capacities of the local communities through (livelihood-related) trainings and by facilitating them to develop their own water management plan. This water management plan was central to a common understanding of the optimal gate operation. A water management forum or platform was also created, involving all important stakeholders of the Khulna area where the community could present their plan and negotiate for some tangible outcome. Capacity needs were assessed in stakeholder workshops and meetings with peri-urban residents through discussion and exchange of ideas.

A capacity development initiative was carried out in Tentultala village in Alutala, because of its peri-urban location and multiple water-related problems. As the most marginalized group, the villagers were the primary recipients of the capacity building interventions. This enabled them to interact and negotiate with decision-makers more effectively and meaningfully in the water management forum. Stakeholder interaction in the forum was important to inform the decision-makers of the peri-urban needs . The village is not under the city administration, but far away from the center of the Union Parishad. Due to its unfavorable location, services from KCC or from the rural governance system (i.e. *Union Parishad*) do not reach the inhabitants of the village. Figure 7.5 gives a closer view of the peri-urban area, where the sluice gate is located. This resource map, prepared by the local community, indicates their main resources of interest such as rivers, canals, fish farms, agricultural land, embankments and roads. Notably, the community highlights the encroached canals, a series of cross-dams placed for fish farms along the abandoned channel of the Mayur River, and the locations of a proposed sluice gate and DTWs.

Tentultala Village is situated approximately ten kilometres away from Khulna city. According to a Bangladesh Bureau of Statistics (2011) census report, there are 668 households in this village. The village depends upon a single water source (river) for daily water demands. The Rupsha and Mayur rivers are flowing on the

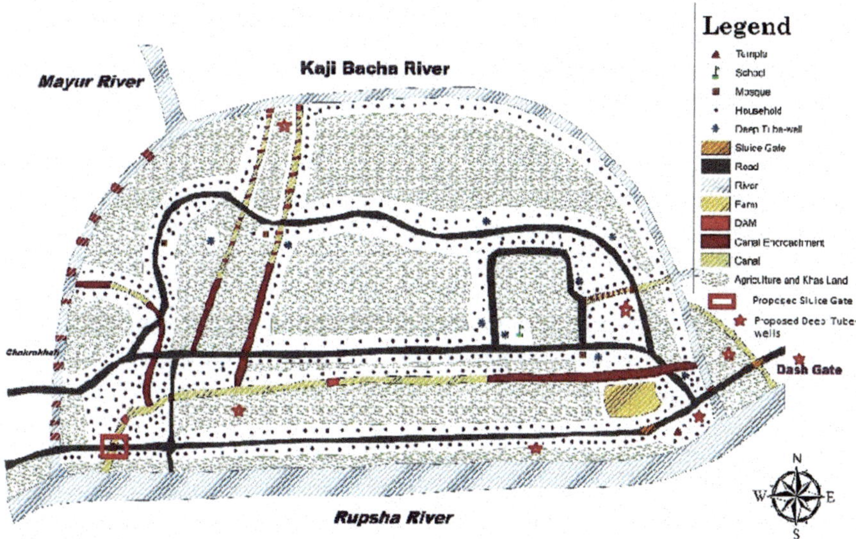

Fig. 7.5 Resource map of areas around the Alutala sluice gate

south-east and north-west sides of the village. As a result, this village is seriously vulnerable to water-related natural disasters like floods. Generally the villagers are engaged in crop cultivation and fish farming as their main and secondary occupation, respectively. Around 80% of the villagers in Tentultala are engaged in cultivating crops like rice, fruits and vegetables. Most farmers do not own land. Crop cultivation started in Tentultala after the establishment of the embankment and the sluice gate. Before construction of the sluice gate, people used to cultivate only one crop in a year (in the *aman* season). Later, villagers could produce three crops each year: rice was produced twice (*robi* season and *aman* season), and vegetables were produced in the winter season. People depend on rainfall for irrigation in the *aman* season. But in the *boro* season they use shallow boring for irrigation.[5] According to the villagers, shallow boring for irrigation purposes is not very effective because of the high iron content of the groundwater.

The main complaint of the Tentultala villagers is that the gate is mainly operated considering the benefits or privileges of other upstream peri-urban communities (e.g. Chak Ahsankhali). The elevation of Tentultala village is 6 feet higher than the upstream areas. Those areas would be flooded if Tentultala villagers were allowed to get sufficient water for their crops. For these reasons, gate operation is mainly conducted considering the privileges of the upstream communities. The fishermen group is also affected by the improper management of the sluice gate, as they cannot get sufficient water for fish farming. Further, water logging is a serious threat for Tentultala. The drainage system in this area is poor; there is no proper drainage system which considers the direction of water flows. The only canal is supposed to be connected with the main river, but usually becomes disconnected from the river due to encroachment and illegal grabbing by powerful political leaders and fish farms. Improper sluice gate management is another cause of drainage problems.

Against this background, collaborative action towards capacity development for securing water in the peri-urban area under the CCMCC project was initiated by the partner organizations of Bangladesh (BUET and JJS). To solve all conflicts, a participatory water management plan and a water forum involving upstream and downstream dwellers, KCC, BWDB, *Union Parishad* and *Upazila Parishad*[6] were considered to be important, and capacity development of the stakeholders for preparation of the plan was attempted. A number of steps were followed to develop the water management plan of Tentultala. Various capacity building initiatives were taken by JJS to enhance the capacity of the people. The project has provided necessary information and technical support to the community for developing the plan. Several discussion sessions were held at the community level, with the participation of both male and female members from various sections. During the process,

---

[5] Aman and boro are rice seasons. Aman rice is generally sown in the rainy season and harvested in winter. Boro rice is sown in winter and harvested before the rainy season.

[6] Sub-district council and the middle tier of the local government system in Bangladesh. An Upazila consists of several Unions.

experts developed a deeper understanding of the conflict dynamics, and more practical understanding of mitigations measures through the stakeholder interactions.

A number of multi-stakeholder platform formation meetings, workshops on negotiation, and cooperation for water management were arranged during the project period to identify the water-related problems of the peri-urban communities, to develop action plans for negotiation with the targeted stakeholders, and to manage water-related conflicts for efficient water management. The people of Tentultala have identified many problems of their area through discussions and exercises. The identified problems are drinking water (tube-wells), water logging and canal encroachment, road communication, electricity, cyclone shelter, community clinic and sanitation. Then they prioritized the issues and identified three major issues related to water: (i) water logging and canal encroachment; (ii) scarcity of safe drinking water; (iii) water pollution of the Mayur/Kazi Bacha rivers. After prioritizing the issues, the community identified the related stakeholders and institutions that can contribute to resolving their issues. The project team created an opportunity for the community to discuss these issues through a stakeholder consultation meeting, where they presented their water management plan. In the consultation meeting, participants gave suggestions to improve the plan. The project then assisted the community to finalize the plan, which consisted of the following main components:

Water logging and canal encroachment: the community planned to remove barriers on the canals and to stop canal encroachment and illegal land grabbing through discussion and the submission of a memorandum to the *Upazila*, District and *Union Parishad* administrations. They initiated advocacy with BWDB to build a sluice gate in the area of *Boro Khal* for draining out excess water in the rainy season as well as for irrigation in the dry season. The people also planned the formulation of a sluice gate management rule for smooth operation of Alutala sluice gate.

Scarcity of safe drinking water: to mitigate the drinking water crisis, the community identified seven sites for installation of DTWs. The community talked with the Department of Public Health Engineering (a government agency responsible for drilling DTWs in rural areas) and *Union Parishad* Chairman for allocating seven DTWs in their proposed area.

Water Pollution of the Mayur/Kazi Bacha River: the Tentultala community has developed a plan for safe disposal of wastes in the Mayur River, to be discussed with KCC. An interface meeting was organized with KCC, DoE and other related organizations to discuss the treatment of clinical, industrial and slaughterhouse wastes before discharging to the Mayur River. The project team created the opportunity for the community to discuss the issues with the relevant authorities.

The process followed in preparing and implementing the water management plan is schematized below (Fig. 7.6):

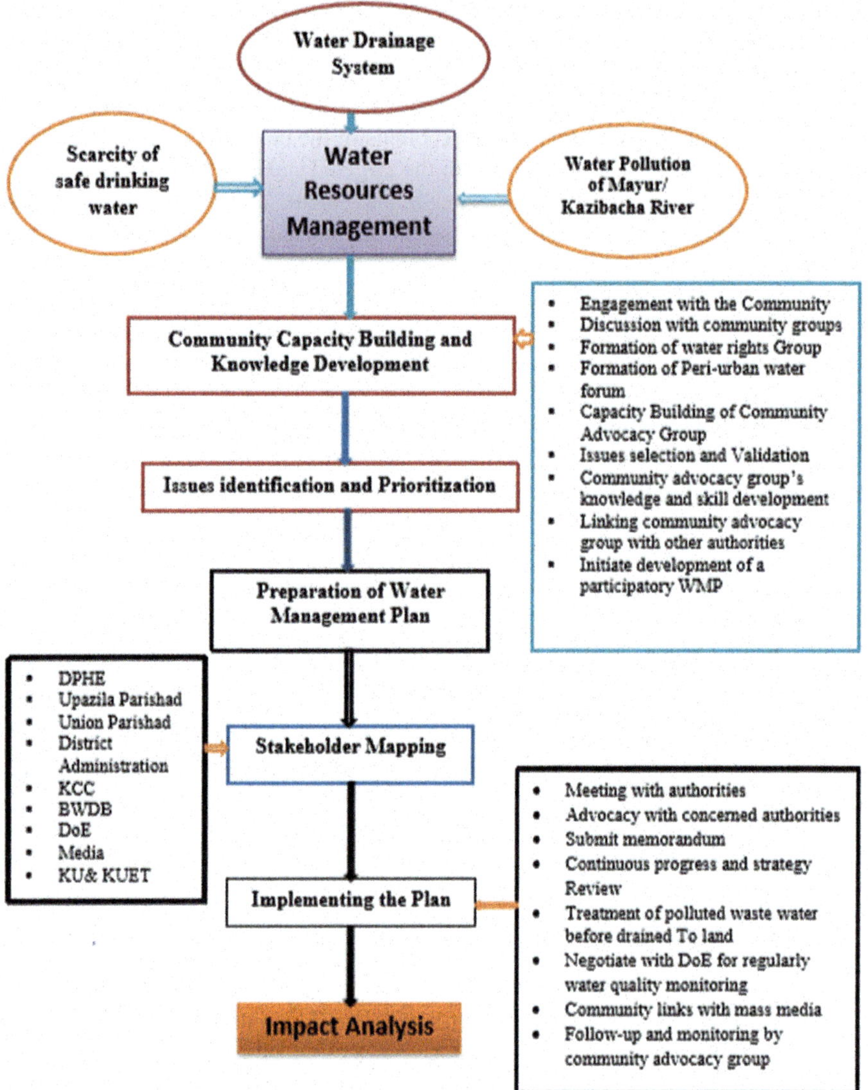

Fig. 7.6 Steps followed in preparing and implementing the small-scale participatory water management plan for Alutala

7.5.3 *Implementing the Plan*

For implementing the plan, some processes were initiated to achieve the desired goal. The community will benefit in the long run if the prepared plan is implemented in a sustainable manner. The processes addressing the three major issues are summarized in Table 7.2.

Table 7.2 Main processes in the implementation plan

Major issues	Process
Water drainage system	Community advocacy group organizes meetings with authorities and departments
	Community conducts advocacy with *Upazila* administration for stopping canal encroachment and illegal land grabbing
	Community conducts advocacy with BWDB to prepare a sluice gatein the area of *Boro Khal* (in front of CSS)
	Community submits memorandum to *upazila* and district administration and *Union Parishad* for stopping canal encroachment and illegal land grabbing
	Continuous progress and strategy review, follow-up and monitoring
Scarcity of safe drinking water	Community advocacy group organizes meetings with authorities and departments
	Community conducts advocacy with district and *Upazila* DPHE and *Union Parishad* for installation of safe drinking water options
	Community people submit application to *Upazila* chairman and DPHE for installation of deep tube-wells
	Continuous progress and strategy review, follow-up and monitoring
Water pollution of the Mayur River	Advocacy group demands treatment of polluted wastewater before drained to the Mayur River
	Advocacy group negotiates with DoE for regular monitoring of the Mayur river water quality
	Community links with mass media on this issue
	Continuous progress and strategy review, follow-up and monitoring

In a Peri-urban Water Forum meeting in Khulna, the village community presented the peri-urban water management plan through a social and resource map of Tentultala, prepared by themselves. In the water management plan the community identified the existing problems of the area and enlisted the responsible department for further discussion. They also fixed a timeline to address the issues. Aside from these, individual stakeholder meetings with DPHE and BWDB were organized to discuss the small scale-participatory water management plan of Tentultala village.

The following outcomes and achievements of the action research are highlighted:

- The *Union Parishad* installed a DTW near the sluice gate;
- The *Upazila* administration took initiatives to remove canal barriers (cross-dams) and evict illegal grabbing;
- KCC appointed a permanent gate operator for operation of the sluice gate;
- Female and male community leadership was developed, to participate in local, regional and national level dialogue.

7.5.4 Legal Remedy

Community Cooperative Fish Farming Project
The commercial fish and shrimp farming in the Mayur River, started by the urban elites in the early 1990s, directly hit the fishermen community of Alutala through eviction from their inherited livelihood. The frustrated fishermen community got organized and established a movement to protest this commercial shrimp farming at the cost of freshwater fisheries. In response, the urban elites filed a case against their protest. Some local NGOs (e.g. *Shomaj Progati Shangha*, Action Aid) and human rights organizations stood by their side with legal support to rehabilitate and re-establish their right to the Mayur River. The NGOs consulted with, and also involved, BWDB and the Department of Fisheries to assist the peri-urban fishermen community. Eventually the case was dismissed in favor of the fishermen community, though the social conflict between the poor fishermen and the local influentials who failed to cultivate shrimp continues to exist there. In response to the unfavorable gate operation, the community started a fish farming project (the concept came from small households) by leasing *khas* land with an informal and unauthorized political approval about 10 years ago. They have officially applied for taking the land on lease with the DC (District Commissioner, the government authority) but have not yet received formal approval. The benefits are small: through this fish farming project, each family gets only BDT 2500 in a year.[7]

River Encroachment
Another legal victory came recently, when the High Court ruled to remove all illegal structures and encroachments along the Mayur River, some of which KCC had already started implementing. This case was filed by a local NGO. From these two instances, it can be said that NGOs as mediators can often empower local communities. Such legal courses take time but ultimately reduce many disputes and provide sustainable resolutions. The research team has been a passive partner in these legal processes. Representatives of various NGOs often took part in our workshops and meetings. We supported their activities and provided moral support through our presence in the field during the last 10 years. We also met with two mayors of KCC during this time period and apprised them of our activities, which were appreciated and encouraged by both of them. Such engagement at various levels provided a moral support to the process of conflict resolution.

7.6 Conclusion

Conflicts around water, livelihoods, and ecosystem services and functions in peri-urban Khulna largely emerged from inequitable design and operation of the sluice gate. These conflicts are expressed in the relationships between the stakeholders and

---

[7] 1 US Dollar = 85 Bangladeshi Taka (BDT)

actors beyond the peri-urban space and vary with time. The original design of the structure prioritized urban requirements while ignoring peri-urban and ecosystem needs. With no provision for ecological and hydrological continuity across the embankment and the gate, the fisheries, agriculture and other associated peri-urban livelihood sources were severely affected. For gate operation as well, urban storm-water and wastewater discharge had a priority over peri-urban water requirements. Later, the degrading water quality of the Mayur River, partly caused by unmanaged urban wastewater and partly by manipulated gate operation, also gave rise to the conflicts in the peri-urban area. Climate change has a clear relation to the water conflicts, which is evident from the additional stress of salinity intrusion induced by sea level rise and high-intensity rainfall events causing water logging. Soil and groundwater salinity has increased with river salinity, affecting agriculture, fisheries and drinking water sources.

An optimal gate operation to meet peri-urban needs is challenged by several factors. First, the peri-urban area is not a formal administrative entity in either urban or rural governance. Thus the unique transient nature of hydro-social processes in the peri-urban space is not addressed in policies or conflict management strategies related to the sluice gate. The peri-urban space is seen as only a contextual extension of either urban or rural issues. The sluice gate was constructed to provide very specific protection and services to the urban area only. The broader and long-term considerations for integrating the social, hydrological, ecological and environmental needs, particularly also for the downstream peri-urban communities, were missing in the technical design. Second, the roles and interests of peri-urban stakeholders around gate operation change as the peri-urban space undergoes transitions. These changes in stakeholder relations often re-align with the social power structure and lead to new forms of conflict and new power alliances. Legal recourse may temporarily settle disputes that are beyond mediation through dialogue and negotiation. The unmitigated intrinsic conflict shifts the stakeholder and power relations to a new equilibrium. Third, urban policy actors and implementing agencies such as KCC and BWDB dominate gate operation. Representation of the peri-urban communities in the decision-making process is nominal and ineffective unless there is a stake of the urban institutions or their representatives in peri-urban issues. Fourth, conflicts between public agencies may worsen peri-urban water insecurity and force the communities to seek alternative means of livelihood. The current minimally-functional state of the gate is a result of its dual responsibility and control: KCC decides only on gate operation while BWDB, despite having no interest in the gate, is still responsible for its repair and maintenance. There is no formal space for coordination or dialogue among these agencies, along with the LGIs and peri-urban communities.

Capacity development of the marginalized peri-urban communities proved to be effective for engaging them in dialogue and negotiation with policy actors and implementing agencies for conflict mitigation around the sluice gate. Such capacities include a shared knowledge of the hydro-ecological system in relation to their lives and livelihoods, stakeholder relations, organizational norms and practices, and preparation of an integrated and strategic water management plan. Sustainability of the engagement of the peri-urban communities with policy actors and stakeholders, and dialogue on a collaborative platform, however, depend much on the active role of a third mediating party, such as a local NGO. The dubious roles of powerful stakeholders like LGIs, who have stakes on both sides of the conflict around the sluice gate, politically empowered locals, and fish farm owners may also undermine the effectiveness of the platform for dialogue and negotiation. Early engagement with key policy actors like the KCC, mayor and LGIs also proved to be helpful to create joint ownership of shared issues and concerns, and to align the multi-stakeholder platform for collaborative action.

We conclude this chapter with the following key learnings from the action research. First, continued advocacy with responsible authorities for successful implementation of a participatory water management plan eventually bears fruit, although the process is time-consuming and requires sustained efforts. Second, strengthening of cooperation between communities, government departments, local administration and local government is essential to solve community issues and problems. Such strengthening can occur through the engagement of a neutral actor, which in this case was our local partner, JJS. Third, ownership of a dialogue platform builds on trust, mutual understanding and respect for each other's positions, not only on negotiations and demands. Close engagement of a local facilitating party trusted by all is crucial in this process and its continuity.

References

Bangladesh Bureau of Statistics. (2011). *Statistical yearbook of Bangladesh*. Government of the People's Republic of Bangladesh.

Boelens, R., Hoogesteger, J., Swyngedouw, E., Vos, J., & Wester, P. (2016). Hydrosocial territories: A political ecology perspective. *Water International, 41*(1), 1–14.

Dasgupta, P., Ghosh, P. K., Rahman, R., & Khan, M. S. A. (2016). Mapping gains and losses from the Mayur ecosystem. In V. Narain & A. Prakash (Eds.), *Water security in peri-urban South Asia: Adapting to climate change and urbanization* (pp. 261–285). Oxford University Press.

Hallegatte, S., Green, C., Nicholls, R. J., & Corfee-Morlot, J. (2013). Future flood losses in major coastal cities. *Nature Climate Change, 3*, 802–806.

Hanson, S., Nicholls, R. J., Ranger, N., Hallegatte, S., Corfee-Morlot, J., Herweijer, C., & Chateau, J. (2011). A global ranking of port cities with high exposure to climate extremes. *Climatic Change, 104*, 89–111.

Khan, M. S. A., Mondal, M. S., Rahman, R., Huq, H., Dutta, D. K., Kumar, U., & Jalal, M. R. (2016). Urban burden on peri-urban areas: Shared use of a river in a coastal city vulnerable to climate change. In V. Narain & A. Prakash (Eds.), *Water security in peri-urban South Asia: Adapting to climate change and urbanization* (pp. 33–74). Oxford University Press.

Khan, M. S. A., Mondal, M. S., Sada, R., & Gummadilli, S. (2013). *Climatic trends and variability in South Asia: A case of four peri-urban locations*. SaciWATERs and IDRC.

Massuel, S., Riaux, J., Molle, F., Kuper, M., Ogilvie, A., Collard, A. L., Leduc, C., & Barreteau, O. (2018). Inspiring a broader socio-hydrological negotiation approach with interdisciplinary field-based experience. *Water Resources Research, 54*(4), 2510–2522.

Mondal, M. S., Jalal, M. R., Khan, M. S. A., Kumar, U., Rahman, R., & Huq, H. (2013). Hydro-meteorological trends in southwest coastal Bangladesh: Perspectives of climate change and human interventions. *American Journal of Climate Change, 2013*(2), 62–70.

Murshed, S. B., & Khan, M. S. A. (2009). Water use conflict between agriculture and fisheries in a selected water resources development project in Bangladesh. *SAWAS, 1*(2), 159–183.

Mutahara, M., Warner, J., & Khan, M. S. A. (2019). Analyzing the coexistence of conflict and cooperation in a regional delta management system: Tidal River Management (TRM) in the Bangladesh delta. *International Journal of Environmental Policy and Governance, 29*(5), 326–343.

Narain, V., Khan, M. S. A., Sada, R., Singh, S., & Prakash, A. (2013). Urbanization, peri-urban water (in)security and human well-being: A perspective from four South Asian cities. *Water International.* International Water Resources Association and Routledge, *38*(7), 930–940.

Roth, D., Khan, M. S. A., Jahan, I., Rahman, R., Narain, V., Singh, A. K., Priya, M., Sen, S., Shrestha, A., & Yakami, S. (2018). Climates of urbanization: Local experiences of water security, conflict and cooperation in peri-urban South-Asia. *Climate Policy, 19*(sup1), S78–S93.

Wesselink, A., Kooy, M., & Warner, J. (2017). Socio-hydrology and hydrosocial analysis: Toward dialogues across disciplines. *WIREs Water, 4*, e1196.

Chapter 8
Interventions to Strengthen Institutional Capacity for Peri-Urban Water Management in South Asia

Sharlene L. Gomes

8.1 The Institutional Context of Peri-Urban Water Problems

In South Asia, urbanization is growing at a rapid pace. The recent "World Urbanization Prospects" report projects India to account for 35% of the global urban population growth between 2018 and 2050 (United Nations, 2018). Cities across the Indian sub-continent are expanding rapidly, increasing the demand for and use of surface water and groundwater. The pressure on local water bodies is especially felt at the urban fringes, also known as the peri-urban space. These are the transition zones during the urbanization process, where the shift from rural to urban is most visible in changing land use, population, economic activities, and institutions (Allen, 2003; Iaquinta & Drescher, 2000; Narain, 2010). Many peri-urban communities in India that were historically agricultural in terms of livelihoods are transitioning to other types of economic activities. This transition process, however, varies across contexts. Peri-urban spaces of Kolkata, for example, have seen the emergence of small-scale industries, in particular, dyeing and bleaching factories, whereas areas near Pune are becoming a hub for horticulture and nurseries.

Water plays an important role in shaping the urban transition process. Peri-urban spaces need water for drinking, domestic, and livelihood-related purposes. The source of water for these diverse uses varies across geographic regions. In rain-fed areas of Maharashtra for example, people get water from surface water bodies controlled by dams. In West Bengal, groundwater is the primary source of water for domestic or livelihood activities such as farming (Fig. 8.1a, b). In drought-prone Telangana, water is often drawn from large tanks referred to locally as *cheruvus*

S. L. Gomes (✉)
Faculty of Technology, Policy, and Management, Delft University of Technology, Delft, The Netherlands
e-mail: S.L.Gomes@tudelft.nl

V. Narain, D. Roth (eds.), *Water Security, Conflict and Cooperation in Peri-Urban South Asia*, https://doi.org/10.1007/978-3-030-79035-6_8

Fig. 8.1 Types of water sources in peri-urban South Asia: (**a**) tube-well in Kolkata, (**b**) well in Pune used for irrigation, and (**c**) *Cheruvu* in Hyderabad (photos Sharlene Gomes, courtesy The Researcher)

(Fig. 8.1c). Managing water in peri-urban spaces is often problematic, and there are growing concerns of water insecurity in this context across South Asia.

Peri-urban water resources are under threat due to growing demand for, and resulting pressures. Livelihood shifts and migration to the urban fringes change the demands for and uses of water. At the same time, nearby cities also struggle to meet the supply needs of urban residents, requiring service providers to find additional water sources further away. This creates competition for water access and sometimes results in conflicts. Studies from peri-urban regions highlight different conflicts over water. In peri-urban Khulna (Bangladesh), water-related conflicts are found in both the domains of domestic and livelihood water uses. Roth et al. (2019) describe issues between commercial (saline) fish farmers and freshwater fishermen over operation of a canal gate in Alutala. Nearby in Phultala, a court case was filed in 2008 by residents against the Khulna city corporation, to safeguard water access from a large-scale groundwater abstraction project (Gomes & Hermans, 2018). Conflicts between peri-urban villages and local industries have also occurred throughout South Asia; for example, locally led campaigns against brick-making factories and water tanker businesses in the Kathmandu valley (Roth et al., 2019), and dyeing factories in peri-urban Kolkata (Gomes, 2019).

The status of water resources in peri-urban South Asia is in part, a result of the quality and functioning of water-related institutions. Institutions may be defined as societal "rules" which guide decision-making and interactions (North, 1990). These rules can be both formal and informal. Formal rules are those that are codified and enforced via laws, policies, and regulations, whereas informal rules are socially transmitted norms, customs, and traditions. Both types of institutions co-exist in peri-urban areas. There is evidence that formal institutions for water management are ineffective in many peri-urban areas. This, in part, stems from how institutions are arranged along rural and urban administrative boundaries, leaving peri-urban areas with unclear, overlapping and often fragmented institutions. Hence, clearly defining the roles and responsibilities for water management within this changing institutional setup proves challenging. Moreover, peri-urban areas are sometimes officially considered part of the rural administration.[1] This governance gap leads to unresolved conflicts and increasing water insecurity (Sen et al., 2017). Rural institutions for water management fail to adequately cater to rapidly changing local needs. Often, peri-urban actors utilize informal solutions to manage daily water needs, particularly if they are more effective or have greater local legitimacy than their formal counterparts. Research from the *Shifting Grounds* project highlights how informal institutions help residents from Hogladanga village (in peri-urban Khulna) access drinking water (Gomes & Hermans, 2018).

The significance of the peri-urban institutional context suggests the importance of considering them during problem-solving and in the design of water management reforms. Yet, institutions are often overlooked. This neglect can stem from limited knowledge of the institutional context. Decision-makers require a local-level understanding of peri-urban water needs and what this means from an institutional perspective. Failure to recognize the variability in hydrological conditions or differences in specific rural-to-urban trajectories, for example, can result in institutions that are locally ineffective or become redundant over time as pressures and needs evolve. Similarly, peri-urban villages might overlook certain institutional solutions if they ignore the institutional context when examining local water problems. This can be expected in peri-urban contexts, where decision-making is often fragmented.

These challenges suggest a need for capacity building at both levels. When the underlying problems in peri-urban water governance relate to institutions, capacity building offers actors the ability to navigate their institutional context during problem solving. In this chapter, two types of structured, participatory approaches are explored, both of which emphasize the institutional elements of problems faced in peri-urban contexts. They offer guidance to researchers and practitioners working with peri-urban actors to strengthen water governance during urbanization.

---

[1] Although the provision of *Nagar Panchayats* (Notified Area Councils) in the 74th Amendment Act (1992) in India, for instance, was meant precisely for transitional spaces like this, it is often not implemented (Shaw, 2005).

8.2 Supporting Peri-Urban Decision-Making Through Capacity Building

A number of contextual features are important to consider when designing interventions for peri-urban areas. First, the uniqueness of the institutional context demonstrates why institutions are important to consider in the framing and analysis of peri-urban dilemmas. Institutions are especially relevant if the prescribed rules, their interpretation, or their implementation lie at the root of the problem. Here, actors' perspectives of the problem may be enhanced by framing them through an institutional lens. In some cases, this helps ensure that actors address the "right" problem, while in others this perspective can open up the solution space incuding the possibility for institutional change.

A second point is the inherent multi-actor nature of this context. The term "peri-urban" by definition refers to transitional spaces, therefore, problems typically involve a diverse set of users and decision-makers. Its social composition also evolves over time. Planning and management of water resources requires taking into consideration the needs and concerns of multiple actors. Here, outcomes are the result of negotiations and trade-offs. Therefore, knowing how to navigate this multi-actor space is valuable. Furthermore, cooperation serves as an alternative strategy to address problems during urban transitions when decision-making is fragmented. Here, problem-solving efforts can benefit from a deeper understanding of the actors, their objectives, resources, and values.

Third, peri-urban water resources exist in a larger system that is both complex and dynamic. In addition to the hydrological complexity of the resource itself, the system is also shaped by the institutions and actors who use and apply them in their role as resource users, providers, managers or regulators. This requires an integrated approach to tackling peri-urban water challenges, recognising the competing needs between actors, different sectors, or across scales. In South Asia, where decision-making is, at times, constrained by politics or department-level operational and financial silos, solutions may require adjusting the underlying rules. Meanwhile, the dynamic properties of the system are associated with the transitional nature of peri-urban contexts. Solutions and plans for today's peri-urban challenges may fall short or, worse, have unintended consequences in the future as these areas evolve. This calls for a longer-term view of the problem and planning for a range of different, plausible futures. These three contextual elements of peri-urban areas are relevant during capacity building.

A number of established and emerging disciplinary theories, methods, and approaches may be used as the starting point for structuring capacity building interventions. Depending on the context, purpose, and targeted beneficiaries, practitioners may draw from fields like action research, operations research, and policy analysis, among others. Each offers approaches and methods to support problem structuring, diagnosis, and planning activities in peri-urban contexts. Their selection and combination depends on the type of peri-urban challenge the intervention is design to address. The following sections demonstrate how Community Operational

Research (COR) and adaptation pathways are used as a basis for two different peri-urban interventions in South Asia.

The interventions discussed below are undertaken as part of two separate transdisciplinary research projects: *Shifting Grounds* and *H2O-T2S* project in urban fringe areas. Both projects are led by an international consortium of partners from academia and practice. The goal in both is to support decision-making during urban transitions through scientific research and participatory stakeholder engagement. While both projects focus on peri-urban water management, the targeted beneficiaries and capacity building needs differ. For this reason, two structured approaches —the APIA and Transitions Pathways— are outlined below. They serve as the starting point for intervening in the institutional context of water-related challenges in peri-urban South Asia.

8.3 APIA: Approach for Participatory Institutional Analysis

To support the community-led institutional change objectives of the *Shifting Grounds* project in the peri-urban Ganges delta, the Approach for Participatory Institutional Analysis (APIA) was developed. Its design draws from Community Operational Research (COR), a sub-discipline of Operations Research. COR applies OR tools and methods to support marginalized communities (Johnson, 2012). It emphasizes a contextual, systems level analysis of societal problems through active stakeholder participation (Gomes et al., 2018). In the urbanizing Ganges delta, peri-urban communities are isolated from decision-making arenas, resulting in a mismatch between local groundwater needs and institutions, and a narrow solution space for the affected communities to select from. Therefore, the APIA was designed and used in the project as a structured approach to help marginalized communities explore groundwater problems and possible solutions through an institutional lens. The intervention undertaken over 4 years (2014–2018) was expected to build capacity for locally-driven institutional change.

8.3.1 Overview and Methods in APIA

APIA consists of four main steps (Fig. 8.2). Step 1 is about problem identification. Here, peri-urban actors identify their most pressing problems and define the boundaries of that problem for further analysis. The criteria for selecting problems that are amenable for further analysis through APIA are threefold: they reflect problems that are complex and dynamic, multi-actor in nature, and have a distinct institutional feature that requires further attention during the problem-solving process. Once a list of problems is generated, actors prioritize them in order of importance, using a simple ranking process. Taking up the most pressing problems in APIA encourages buy-in during the intervention. Thereafter, the challenge is to define problem

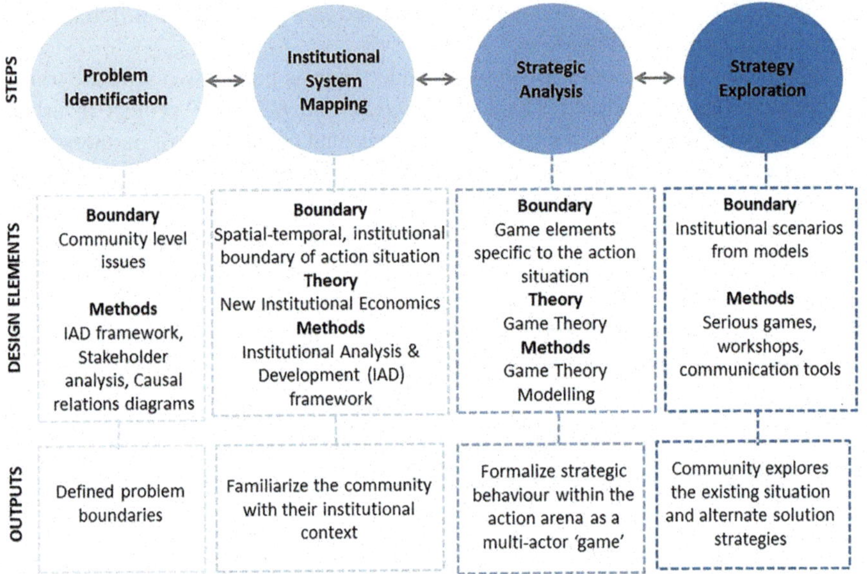

Fig. 8.2 Overview of the APIA (Gomes, 2019)

boundaries. To structure this discussion, methods like stakeholder analysis, social maps, and simple mind-maps or causal diagrams may be used to define the actor, geographic boundaries, and causal relations respectively in the problem.

In step 2, the institutional context of the selected problem is mapped and analysed. For this, the Institutional Analysis and Development (IAD) framework by Ostrom (2005) is a useful guide for the data collection and mapping exercise. The IAD framework is rooted in theories from the field of institutional economics and draws conceptual linkages between institutional and other important contextual features, the multi-actor context (referred to as the action arena), and the outcomes of the problem. Feedback loops in the framework also offer a dynamic view of the system. They help create snapshots of peri-urban issues over time. The desired outcome of step 2 is to familiarize users with the institutions underlying their problem, not simply as stand-alone aspects of the system but in terms of how these institutions are operationalized by actors during decision-making, interactions, and resulting outcomes.

From this larger overview of institutional features in step 2, step 3 zooms in to the multi-actor interactions within the problem. Game theory models are used to structure and analyse actor behavior as a game consisting of actors (or players), actions (or moves), outcomes, and payoffs (or utility) (Rasmusen, 2007). This method is applied in step 3. Actors in the game have a set of actions to choose from depending on the situation they face. The combination of actions by different actors produces an outcome. Outcomes offer some utility to the actors, however, some outcomes are better than others in this regard. In this way, the solution space of the problem is analysed. Although institutions are not an explicit input in a game theory model, they define the "rules of the game". In this way, institutions provide the rules for players in the game, their actions, and overall structure of the game. In step 3, different types of game theory models can be constructed depending on the type of problem being examined. Non-cooperative game theory models are used in situations with fixed rules and where actors pursue individual interests, while cooperative game theory models explore situations where there is a willingness for coordination and sharing of resources (Hermans & Cunningham, 2018; Madani, 2010; Slinger et al., 2014).

Step 4 builds on the results of step 3. Strategy exploration is meant to give users an interactive, immersive experience in the solution-finding process. To facilitate this, game theory models are translated into role-playing games. Gaming-simulation methods simulate real-world phenomena in the roles, rules, and incentives of a game (Meijer & Hofstede, 2003, cited in Meijer, 2009). Game design handbooks call for designs to be based on the purpose and the context in which it is used (Duke, 1980; Greenblat, 1988). A simple role-playing game facilitated with the use of a game board is sufficient. Peri-urban actors take on the role of one of the players in the game. Through role-play, participants experience other actors' decision-making processes in different problem scenarios. Multiple rounds are played to allow for the comparison of outcomes from different strategies. At this stage of APIA, the goal is to build capacity on several fronts: first, knowledge about the other actors in the problem, their values, resources, and strategic preferences; second, potential solutions to the problem both in the short and longer term; third, building skills to negotiate these solutions in the real-world.

8.3.2   Piloting the APIA in Peri-Urban Khulna

Khulna is the third largest city in Bangladesh. Peri-urban areas of this city are extremely dependent on groundwater for almost all purposes. There is evidence of groundwater scarcity, contamination with iron, salinity etc. as a result of overexploitation. One such peri-urban location affected by the groundwater crisis is Hogladanga village, situated approximately 7 kms away from Khulna. Located close to the expanding city, the village is isolated from the rural administration that is responsible for water supply services. Officially, Hogladanga falls under the administrative jurisdiction of the Bhotiagata sub-district or *upazilla*. Lack of access to policy-makers and decision-makers at this level limits the community's ability to effectively address the groundwater issues it faces. To support problem-solving efforts in this village, an APIA-based capacity building intervention was undertaken by partners in the *Shifting Grounds* project. Through an improved problem understanding and platform to engage with local decision-makers, community capacity to intervene institutionally in groundwater management problems was expected to improve. With the help of a local partner based in Khulna, structured stakeholder engagements were undertaken over 4 years in the form of site visits, mango tree meetings (regular, informal meetings in the village with village residents) and workshops (formal, multi-stakeholder meetings in Khulna city).

The intervention began with step 1. Hogladanga residents came up with a list of concerns. Although the project focused on groundwater, the types of problems identified included surface water issues, livelihood insecurity, and waste management among others. Ultimately, locals identified access to safe drinking water supply as a priority concern for their village. Groundwater forms a significant part of this problem, as the village continues to rely on tube-wells for accessing drinking water and no surface water options are available. Further discussions on their negotiation strategy revealed that the problem comprises two aspects: availability of drinking water infrastructure and drinking water quality. A social map (Fig. 8.3) was also prepared with the help of local facilitators from Jagrata Juba Shangha (JJS) to demonstrate the extent of the problem.

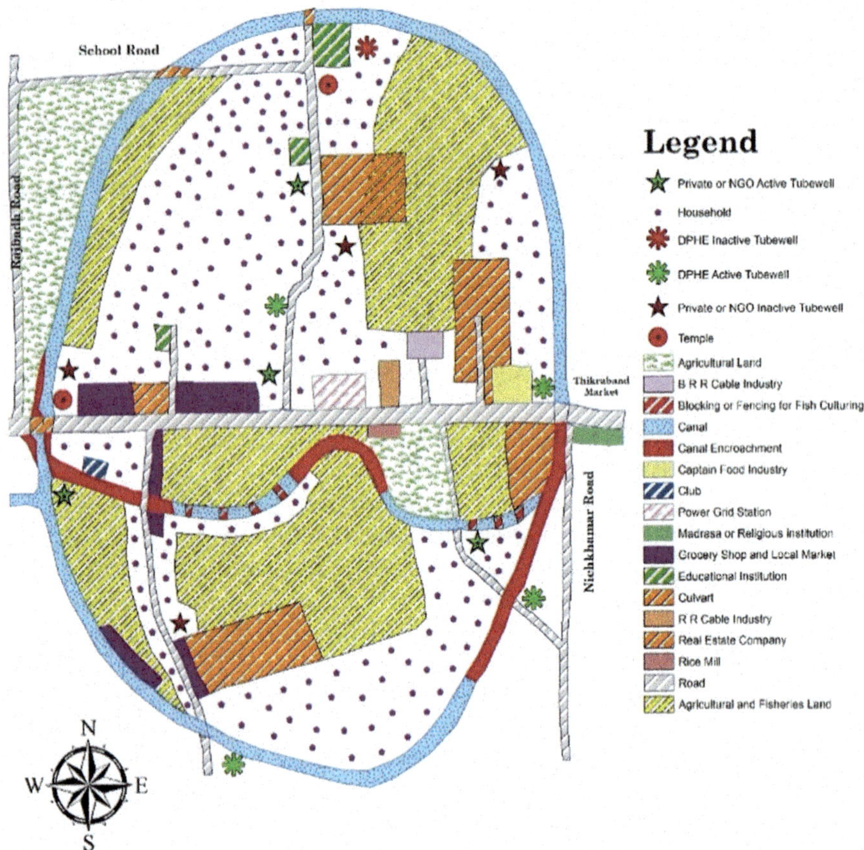

Fig. 8.3 Social map of drinking water supply in Hogladanga. (Image courtesy Jagrata Juba Shangha 2017)

In step 2, the project team (comprising the author and the local partner, JJS) worked with the community to map the institutional context of the drinking water problem. Institutions briefs explaining the formal rules for water supply in Bangladesh were developed and shared with the residents. This was necessary as the village population's understanding of formal rules was limited. The brief contained infographics and short descriptions of key laws, policies and guidelines to supplement discussions in step 2. Subsequently, a workshop was conducted to discuss the findings of institutional mapping. Step 2 highlighted the different phases of the problem. The first phase refers to the initial drinking water problem. It offered insight into why formal rural institutions failed to supply adequate tube-wells to meet the needs of the community. This was due to limited resources for drinking water infrastructure, high demand for public tube-wells, and the politics of distributing tube-well licenses, all leading to an outcome of water insecurity. This was also because, over time, groundwater levels began to decline and tube-wells that had been installed at 400 - 450 ft. depth no longer supplied good quality drinking water. Thus, residents were in need of tube-wells at a depth of 1000 - 1500 ft.

To cope with this, residents looked for informal solutions, thus entering the second (current) phase of the problem. There exists a practice of tube-well sharing between households. Those who could afford it began investing in private tube-wells. These tube-wells are also shared with neighbouring residents. This acts as a makeshift solution to access groundwater for drinking purposes, while residents continue to apply for additional public tube-wells alongside.

In the near future, a third phase of the drinking water problem is expected. A proposal to expand the urban boundaries of Khulna city could bring peri-urban villages into the urban jurisdiction. When this happens, formal institutional arrangements will change from rural to urban, as will the options for drinking water supply, as Khulna city is already supplied with treated surface water via pipelines. Public tube-wells would likely continue to be used in some urban areas in this scenario. The quality, affordability, convenience and reliability varies between these options. Moreover, it was unclear whether water supply projects that were being constructed at the time would cater to the needs of the future urban population. As a result, informal options like private tube-wells and bottled water may still be relied upon in the future.

In step 3, the project team developed game theory models of strategic behaviour using inputs from village residents and other actors. Three models were constructed, each corresponding to a different action arena in the problem. A non-cooperative model represented the existing situation for accessing drinking water infrastructure. A second non-cooperative model represented the future drinking water situation. The collection of improved groundwater data was considered as the first step in addressing the drinking water quality problem, as decisions on the installation of tube-wells were constrained by a lack of aquifer understanding. Hence, a third cooperative model was set up to explore the potential of cooperative strategies to improve groundwater monitoring.

The three game theory models offered several analytical insights for the researcher into the understanding of why actors behave the way they do in drinking water problems, as well as potential solution strategies to address them. However, the models on their own were not easy to interpret by residents from Hogladanga, who lacked a basic understanding of game theory. Representing payoffs as numerical values, for example, paints an abstract picture of a very real problem. Earlier stakeholder discussions in step 3 had expected this, as model inputs like values were difficult to discuss with many residents. Therefore, model results were not presented to the community after step 3. Instead, each model was translated into a role-playing game, as a medium of communication in step 4.

Strategy exploration in step 4 was conducted through an interactive gaming-simulation workshop with Hogladanga residents. Three game sessions were conducted during the workshop. They were facilitated with a game board with movable pieces that was customized in each game. Players were given materials such as role description cards, action cards, resource cards, and scorecards to evaluate outcomes. Participants took on the role of one of the players in the game and were asked to explore the drinking water problem through role play using the materials provided.

In game 1, participants explored solutions to access drinking water supply in the current (peri-urban) scenario. In game 2, they examined the same problem in the future (urban scenario). In both game sessions, players used their resources to make actions, the combinations of which produced an outcome that was then evaluated by scoring their satisfaction level against their players' values (Fig. 8.4). Three rounds were played in both games, allowing participants to explore different strategies to address the drinking water problem. In game 3, participants compared the outcomes of cooperative groundwater monitoring with the round where all players monitored

Fig. 8.4 Strategy exploration through role-play in step 4. (Photo Sharlene Gomes)

the resource individually. In this game, players used their resources to negotiate cooperative agreements. Two rounds were played and outcomes were similarly evaluated against their players' values.[2]

8.3.3 Impact of the APIA in Peri-Urban Khulna

Evaluating the impact of the APIA in Hogladanga was done at each step of the intervention. The direct effects on the community's problem understanding, methods, and facilitation were the evaluation criteria. The intervention highlighted the complexity of the problem. Drinking water insecurity had both infrastructural and quality aspects and, furthermore, changed over time due to urbanization. Understanding the existing situation served as a baseline for comparing alternate scenarios. Institutional mapping built capacity on two levels. First, on the level of the formal institutions that related to water management in general and drinking water supply more specifically. Second, the IAD analysis demonstrated how both formal and informal institutions were operationalized during decision-making and the actors who were involved in the problem over time. Some residents from the village highlighted that the future scenario was most valuable as they had little knowledge of who and what kinds of water supply services would be available.

Further, the community recognized the limits of their solution-finding abilities, given their lack of access to the sub-district administration. Still, APIA findings helped understand the reasons why their preferred outcome may not be feasible, while allowing for a comparison of alternatives. In step 4, the role-playing exercises offered more specific details about the other actors involved in the problem, some of which were previously unknown. Although the intervention did not lead to a new solution, analytical abilities with regard to the solution space improved. For example, in step 4, participants commented on how different strategies require negotiating with different actors and moreover, that these strategies are valued differently by each actor. Negoation experience was also built during the cooperative game. Participants commented on the challenges of negotiating with certain actors, realizing their own role in shaping solutions. This was important, as the community would have to sustain engagements with the government beyond the project.

In terms of methods, visual forms of engagement were positively evaluated in each step. For example, in step 2, while some residents found it difficult to interpret the contents of the institutions brief, the infographic was easily understood. Similarly in step 4, the game board and outcome evaluation scorecards were appreciated. Despite the time needed to initially familiarize with game materials, participants enjoyed the visual and interactive medium of the games itself.

---

[2] For a more detailed account of the strategy exploration workshop, see Gomes (2019).

The role of the local facilitators was critical in the implementation of APIA for different reasons. First, as they were based in Khulna, they had a good local network to support the research activities in each step and were also able to meet regularly with the community. As many as 11 meetings with local households were held over the course of the project. Second, discussing problems of a somewhat sensitive nature was possible thanks to the initial time spent by facilitators to build relationships with the community and convey the goals of the project. Third, this intervention also gave facilitators new skills, as each step of the APIA were impemented with their help in the local language. Since rhis project, JJS has applied the APIA in other projects in Bangladesh (Hossain et al., 2019). Having the same team of facilitators throughout the intervention brought a sense of ease to the discussions. It was observed how women, who were initially reserved, actively engaged in the game sessions and took part in some of the multi-stakeholder dialogues with local government representatives.

Overall, the APIA was a good fit for the type of problems experienced in Hogladanga. Their objectives for participating in this intervention were clear: they wanted a solution to their drinking water problem. While the *Shifting Grounds* project offered a platform to engage regularly with local decision-makers, the APIA offered them knowledge and skills to guide these discussions. In this context, institutional capacity building was achieved through this APIA-based intervention, although wider impacts are yet to be seen.

8.4 Transformative Pathways for Sustainable Peri-Urban Water Management

Policy planning for water management requires a new paradigm, given the uncertainty regarding the future, an important aspect that traditional static plans fail to adequately prepare for (Haasnoot et al., 2013). An adaptive approach that is future-oriented guides decision-making by predicting system changes over time and making strategic plans to manage them. In this way, it combines short-term action with longer-term goals. This is the objective of the Adaptive Pathways approach. Haasnoot et al. (2013) show evidence of this approach being applied to water management in New York, the Rhine and Dutch Deltas, and the Thames estuary. While climate change has been the driving force behind many applications to date, its use in helping peri-urban areas cope with the uncertainty of urbanization had not been explored. Sources of uncertainty in peri-urban water management include anthropogenic factors (such as population, land use change, and economic activities), climatic factors, as well as institutional factors. For example, water resources may be affected differently depending on whether peri-urban areas remain under rural administration, form a *Nagar panchayat*[1] or a new municipality, or get absorbed by a large metropolitan city. Such a future offers both threats and opportunities to water governance. Failing to adequately prepare for the future means that decision-makers will continue relying on ad-hoc coping strategies if today's management plans fail to hold up over time. This can hamper the sustainability and resilience of urban transitions.

To support a more adaptive approach to peri-urban water management, the adaptation pathways approach may be used to structure policy interventions. The *H2O-T2S in Urban Fringe Areas* project does this through the lens of vulnerability and resilience. The project is executed in peri-urban areas of three metropolitan Indian cities: Pune, Hyderabad, and Kolkata. Two peri-urban villages are selected in each context. Through field research and local stakeholder workshops, context-specific vulnerabilities in peri-urban water resources are identified. Baseline studies examine the existing adaptive capacity in each case study to respond to opportunities and threats. These preparatory research activities look at three dimensions in an integrated way: institutions and governance, domestic water access, and livelihood water uses.

Distinctive past and current transformative pathways provide the starting point for exploring future scenarios (Sen et al., 2017). For this, multi-stakeholder workshops are planned as the next step in the project wherein decision-makers, local community representatives, and other experts co-design transformative pathways. The intervention is expected to help peri-urban areas, through their institutional context, cope with short-term vulnerabilities while also stimulating resilience in the longer term (Sen et al., 2017). This forms the intervention phase of the project. These workshops offer a new conceptual approach to govern rural to urban transitions. Local stakeholders in the three case study areas build on the status quo to work towards more sustainable pathways for the future (Sen et al., 2017). In this way, the approach is tailored to the local context.

The three-year project (which began in late 2018) was in progress at the time this book was published. Therefore, it is not possible to demonstrate the results of this approach in peri-urban contexts. Instead, the following section outlines the main steps in the design of transformative pathways. Further, it briefly describes the unique vulnerabilities faced by peri-urban regions around India to highlight key factors to consider during the intervention.

8.4.1 Methodology for Designing Transformative Pathways

Conceptualizing transformative pathways for peri-urban water management makes use of the adaptation pathways approach. Several articles and application manuals outline different stages in the preparation and design of pathways (Bosomworth & Gaillard, 2019; Butler et al., 2016; Coulter, 2019; Haasnoot et al., 2013). For the peri-urban context, these have been adapted as follows:

Stage 1: Define Goals and Objectives
Peri-urban actors start by formulating a clear vision and goals with regard to water management in the future. They define this objective based on what is important for

those involved, their key concerns etc. Here, representation of key peri-urban stakeholders is important, as is clarifying the scope for this objective. On the one hand, having a broad scope is relevant in peri-urban contexts, given the complex, dynamic, and multi-scale characteristics. On the other hand, focusing on a specific type of water management issue or on a particular region keeps adaptation planning exercises manageable for first-time users and generates more context-specific vulnerabilities, future options etc. For example, the opening session of the project conducted by Butler et al. (2016) in Indonesia describes the geographic focus in the intervention in administrative terms (e.g. provincial or sub-district level).

Stage 2: Explore the Current Situation
The current situation refers to drivers of change as it relates to water resources. During this stage, workshop participants describe their operational practices used for peri-urban water management around three aspects: institutions, domestic and livelihood water uses. Here, drivers refer to the underlying causes of issues that concern to peri-urban actors (Butler et al., 2016). This exercise is meant to understand the types of peri-urban vulnerabilities and adaptive capacity that already exists within the system. Facilitators should stress the value of different sources of data and perspectives of the situation, to avoid discounting local and traditional knowledge. Research findings by the project team also serve as inputs during the discussions in stage 2.

Stage 3: Analyse Possible Future Scenarios
Participants describe possible futures for peri-urban water management. Considering a wide range of future scenarios is the strength of the approach. Their desired vision for the future must feature in the list of futures and could even be used as the starting point for discussions regarding alternatives. The remaining futures may be ranked according to how they meet the goals and objectives of peri-urban stakeholders. The business-as-usual, best, and worse-case scenarios may also be identified. It is useful to visualize each scenario using pictures or name them for easier discussion in the subsequent stages of the intervention (see e.g. Butler et al., 2016; Vervoort et al., 2014).

Stage 4: Design Pathways
The design of adaptation pathways can be done by forecasting from the present or back-casting from shared desired future goals (Vervoort et al., 2014). In the forecasting approach, the more commonly used of the two, users begin by identifying actions to address the existing drivers of vulnerability. For each action, a tipping point is identified for that action based on a future system condition. These tipping points serve as triggers for decision-making. One way of identifying potential tipping points is by combining information from current and future scenarios (Coulter, 2019). They explain that a tipping point may have negative consequences (e.g. actions are no longer effective, the point of no return) or may be positive, creating opportunities (e.g. funding, leadership changes). Before reaching this point, peri-urban actions will need to shift to an alternative course of action better equipped for that future condition. For this, a suitable trigger is needed. The type of monitoring

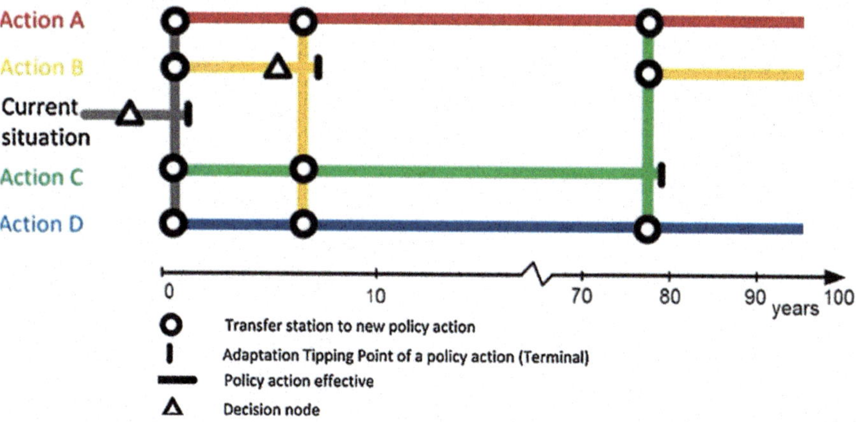

Fig. 8.5 Example of an adaptation pathways schematic. (Adapted from Haasnoot et al., 2013)

and capacity to enact defines how far in advance these triggers are needed (Coulter, 2019; Haasnoot et al., 2013). Next, users identify actions that are robust against a range of possible futures. Promising responses can be clustered together to form the basis of transformation strategy and thereafter, arranged into logical sequences over time, resulting in transformative pathways (Sen et al., 2017). A commonly used schematic for visualizing adaptation pathways is that of a metro map (Fig. 8.5).

Stage 5: Evaluate Pathways Schematic
This helps compare between responses over time and against the goals of the stakeholders involved. Here, trade-offs may be needed with those who bear the cost from that option or have a preference for a more desirable one. Scorecards may be used during the evaluation stages as shown below (Fig. 8.6). This leads to an identification of preferred adaptation pathways.

8.4.2 Water Related Vulnerabilities in Peri-Urban India

Since the project's inception in 2018, a selection of six peri-urban case studies across 3 Indian cities was made. They include the villages of Paud and Uruli Kanchan (Pune), Anajpur and Bowrampet (Hyderabad), Badai and Hadia (Kolkata). Thereafter, the H2O-T2S project team conducted preparatory research activities in the form of site visits, key informant interviews, and focus group discussions with peri-urban stakeholders in each site and government representatives from the local up to the state government level. This offers an initial impression of water-related vulnerabilities across peri-urban locations.

The three regions are shaped by different agro-climatic and hydrogeological conditions, creating differences in resource environments. Moreover, as water is a state subject in India, selecting sites in different Indian states allows for a comparison of

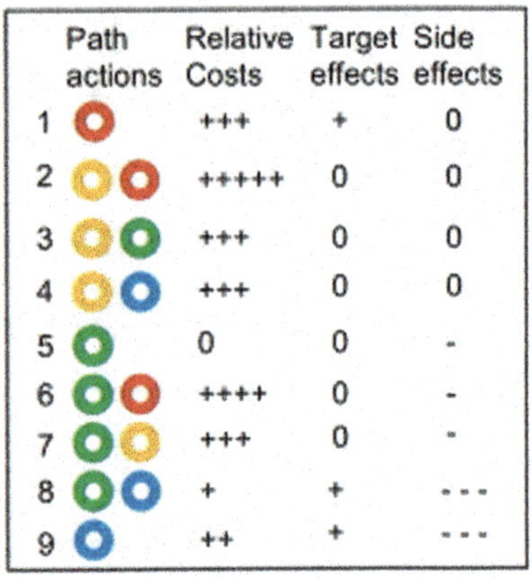

Fig. 8.6 Example of a scorecard used to evaluate adaptation pathways (Haasnoot et al., 2013)

Path actions	Relative Costs	Target effects	Side effects
1	+++	+	0
2	+++++	0	0
3	+++	0	0
4	+++	0	0
5	0	0	-
6	++++	0	-
7	+++	0	-
8	+	+	- - -
9	++	+	- - -

institutional contexts as well. As a result, water dependencies and urban transformations differ significantly across and even within states. For example, major economic activities in Badai and Hadia village, both situated in peri-urban Kolkata, include dyeing factories and wastewater aquaculture respectively. In Badai, this stems from the access to affordable groundwater needed in the manufacturing process. Meanwhile in Hadia, situated in a wetland region, fishermen have access to a large wastewater canal from Kolkata city for fish cultivation. Although both types of economic activities benefit from the access to markets in Kolkata, the need for different water resources as inputs shape their unique economic activities.

Not surprisingly, peri-urban vulnerabilities across India were also context-specific. Paud, situated near Pune in the western Ghats in Maharashtra, has access to surface water resources from large nearby dams. In contrast, Anajpur and Bowrampet near Hyderabad face water shortages as they are situated in drought-prone regions. Furthermore, in Kolkata, some peri-urban areas where local aquifers are contaminated with arsenic or salinity face water insecurity. However, vulnerability does not stem solely from the biophysical environment. Water demand for one use is seen to have knock-on effects elsewhere in the community. For example, small factories are blamed for the release of raw effluents into local surface water bodies. As a result, local farmlands in Badai have, in the past, become contaminated by the release of wastewater from nearby dyeing factories. This resulted in agricultural losses for some peri-urban farmers. Historically, such tensions are also found in Anajpur village between surrounding industries and local farmers, poultry farmers and cattle herders. Meanwhile in Paud, ongoing construction of a large housing complex is expected to nearly double the population and, with it, cause a drastic increase in domestic water demand. The developer plans to invest in its own private drinking water supply, athough separate from the village. As a result, unequal water access could become a problem in the near future.

The government agencies for water management are also sources of water-related vulnerabilities. In West Bengal, regulation of groundwater abstraction is the responsibility of the State Water Investigation Development (SWID), as stated in the Groundwater Resources Act (2005). However, implementation capacity at the block level is limited, as is the sanctioning power to enforce harsher penalties for non-compliance. Meanwhile, interesting institutional set-ups are found in Hyderabad, where urban development centres around the Outer Ring Road (ORR) which surrounds Hyderabad and its peri-urban areas. Drinking water supply within the Hyderabad Municipal area is the responsibility of the Hyderabad Metropolitan Water Supply and Sewerage Board (HMWSSB). Earlier, areas outside Hyderabad city fell under the responsibility of the state government's Rural Water Supply and Sewerage (RWSS), but jurisdiction of the urban service provider was extended to include all areas within the ORR. Therefore, villages within the ORR also receive water from HMWSSB. This example shows how infrastructure to improve urban transportation has also affected the rules for water service provision.

Given these changing dynamics, evidence of adaptation to some of the above mentioned vulnerabilities was observed across the three study regions. Villages like Hadia (Kolkata) and Paud (Pune) are home to cooperative fishing communities (Fig. 8.7). In Paud, the Bhoi fisherfolk are a traditional fishing community, dating back several generations. In Hadia, a large fishing cooperative was formed in 1999. To safeguard their livelihoods from changing dynamics, fishers' co-operatives were established, offering several benefits to members like shared labour responsibilities, access to financial resources for licences, fishing inputs, or emergencies, and risk reduction by sharing profits from fish catch.

With regard to drinking water, the informal sector has helped peri-urban households close the supply gap. Informal providers operate in many peri-urban areas. Small bottled water industries that sell packaged groundwater to neighbouring

Fig. 8.7 Traditional practices of the Bhoi fishing cooperatives. (Photo courtesy C., Butsch)

Fig. 8.8 Informal drinking water providers across peri-urban India: (**a**) water tankers (left); (**b**) RO plants (right). (Photos Sharlene Gomes)

villages, for instance, are mushrooming in parts of peri-urban Kolkata. Although many companies claim that the water is filtered, there are concerns that their illegal nature raises questions of the treatment processes and drinking water standards followed. Elsewhere, in peri-urban Pune and Hyderabad, private and NGO funded Reverse Osmosis (RO) plants can be found besides the public RO plants set up by local *panchayats*.[3] Private tankers also supply water to peri-urban households and businesses, especially during the dry season, and to larger residential complexes (Fig. 8.8).

8.4.3 Designing Transformative Pathways Workshop

Given the status quo of water-related vulnerabilities and peri-urban adaptation strategies, the next step is to design suitable workshops for peri-urban actors to explore transformative strategies. For this, a number of design choices must be considered. An appropriate level for the intervention needs to be selected. While the *H2O-T2S* project planned for state-level workshops, this choice highlights both pros and cons. On the one hand, it enables a focus on state-level policies to sustainably manage uncertain peri-urban futures. Given the fact that water management is shaped to a large extent by state-level policies, it makes sense to target state-level decision-makers when advocating for a more integrated and resilient approach to peri-urban governance.

On the other hand, representation of all stakeholders' views and inputs is essential in the adaptation pathways approach. To ensure this, state-level decision-makers should also participate. Although the research team has a good local network, the availability and willingness of senior government representatives to attend workshops conducted by research projects can prove challenging. Efforts to familiarize

---

[3] *Panchayat* is the most local government administrative level in rural India.

state-level officials with the project, and what it offers in terms of capacity and data, might be needed earlier on in the intervention. Regular updates from the project team would be another way to strengthen relationships before the workshop phase of the project. Furthermore, participation of local communities is also essential. Meaningful and open discussions between government and peri-urban residents can be difficult, given the power differences between these actors. The project will consider ways to manage this in the design of the workshops.

The preparatory assessment presented in the previous section highlights that, even within the same metropolitan area, significant differences in vulnerability and coping capacities can be found between (and even within) peri-urban villages. Therefore, one should consider how state-level workshops can design pathways that are context-specific. It may be necessary to focus on issues common across case-studies from that region or conduct pathways development exercises in parallel within two smaller groups of participants. Another option would be to begin the pathways exercise in a local-level workshop, with communities and local decision-makers. Once their vulnerabilities and visions for the future have been incorporated, a state level workshop will continue the design of pathways.

The three project regions offer sufficient differences for knowledge sharing in the national-level workshop that is planned at the end of the intervention. This will help put peri-urban governance on the national agenda. Moreover, national-level policymakers will hear directly from the state and local-level actors about their context-specific pathways for managing water-related vulnerabilities during urban transitions.

While there is sufficient literature available on the stages in the adaptation pathways approach, the methods used to facilitate this with stakeholders are less clear. Little has been written about how the approach is applied or tailored to specific application contexts. Co-designing transformative pathways with stakeholders in the *H2O-T2S* project is expected to generate insights on its use as a tool for planning. The experiences gained from multiple project sites are critical. They will contribute necessary improvements to this innovative approach for capacity building, to benefit future uses elsewhere (Sen et al., 2017).

8.5 Institutional Reforms in Peri-Urban South Asia: Ways Forward

In this chapter, I have discussed the important role that peri-urban institutions play in water issues during urban transitions, highlighting a need for capacity building on this particular aspect of decision-making. Two different approaches to capacity building are offered: the APIA and the use of transformative pathways for sustainable peri-urban water management. These approaches may be used to design and structure interventions in peri-urban contexts. Both approaches allow for peri-urban actors to explore decision-making problems through an institutional lens, albeit with different objectives. The APIA helps with problem diagnosis and strategy exploration during problem solving. In Bangladesh, this was used to help marginalized communities engage in decision-making arenas while addressing local concerns. Transformative pathways, based on the adaptation pathways approach to planning, allow for actors to explore longer-term adaptive plans, given the uncertainty of peri-urban futures.

Structuring interventions in the institutional context of peri-urban South Asia would benefit from these and other similar participatory approaches. Yet more work is needed to further improve, test, and evaluate these methods in real-world applications. While the transformative pathways intervention in the *H2O-T2S* project is still in an early stage, further research on the APIA is currently underway. Further testing of APIA was also conducted in peri-urban Kolkata. Since the *Shifting Grounds* project, a practical manual for APIA has also been developed (Gomes, 2020). This is meant to expand the application potential of APIA beyond peri-urban contexts and marginalized communities. The manual builds on the lessons from the pilot application in Hogladanga by revising the steps of APIA to give users even more ownership in the capacity building process. In this way, users will have the ability to apply the methods of the APIA themselves, while project partners take on a supporting or facilitating role in the process. Application of APIA manual with professionals is currently underway in Bangladesh to train water professions and examine different types of water problems in urbanizing deltas (Hossain et al., 2019).

Reflection on the above two capacity-building interventions to strengthen peri-urban institutions highlights a number of lessons learned and areas for further research. At the outset, it is important to highlight the fact that institutional change is a slow moving process that is not likely to occur within the timelines of a typical research project. Therefore, a meaningful starting point is to focus on a new conceptual framing of peri-urban contexts and its water-related challenges. In many cases, addressing a gap in knowledge or creating a platform to engage with other peri-urban stakeholders is often a necessary and useful first step.

Furthermore, the transferability of these approaches to address different types of problems facing peri-urban actors needs to be explored. APIA, for example, has been evaluated in other peri-urban contexts with other types of actors (e.g. government) and for other types of problems in Kolkata (Gomes, 2019). This helped understand their value and application potential of the APIA in peri-urban contexts, beyond those they were initially designed to support. Similarly, there may be other tools (see e.g. Ducrot, 2009) that may also be potentially useful for structuring peri-urban interventions. A full scan of approaches available in the literature is, however, not part of this chapter.

The timing of these interventions is also relevant. Despite the availability of suitable tools, resources and networks, activities may be influenced by conditions present at the time. Political changes during an intervention can slow or even hamper activities if elected officials are not supportive. Peri-urban areas also deal with problems like inequity as a result of caste differences or corrupt practices. Projects need to be aware of these and other sensitive topics during interventions to ensure that stakeholders do not face any negative consequences from participating and feel comfortable discussing them during activities.

Facilitation is also extremely important for the success of these interventions. It is useful to have local facilitators, who already have good relationships and networks in the context and are able to conduct the steps in the local language. The results from the *Shifting Grounds* project reflect this. In *H2O-T2S*, local partners represent each of the three metropolitan regions. Nevertheless, the ability to meaningfully engage with senior state and national actors remains to be seen.

The two interventions described in this chapter demonstrate ways in which peri-urban areas may be helped. Through support in decision-making, problem understanding and planning, these interventions buid peri-urban stakeholders' capacity to manage their complex, dynamic and uncertain context. The approaches in this chapter present ways of intervening in peri-urban dilemmas with conflicting needs and objectives, through improvements in longer-term planning and more focused problem-solving interventions. It is through a better understanding of the challenges faced in peri-urban contexts that sustainable water governance during urban transitions can be achieved.

Acknowledgements The project *H2O-T2S in Urban Fringe Areas* is financially supported by the Belmont Forum and NORFACE Joint Research Programme on Transformations to Sustainability, which is co-funded by AKA, ANR, DLR/BMBF, ESRC, FAPESP, FNRS, FWO, ISSC, JST, NSF, NWO, RCN, VR, and the European Commission through Horizon 2020 under grant agreement No. 730211. The *Shifting Grounds* project was funded by the Urbanising Deltas of the World Programme of the Netherlands Organisation for Scientific Research (NWO) under grant No. W 07.69.104.

References

Allen, A. (2003). Environmental planning and management of the peri-urban interface: Perspectives on an emerging field. *Environment and Urbanization, 15*(1), 135–148.

Bosomworth, K., & Gaillard, E. (2019). Engaging with uncertainty *and* ambiguity through participatory 'adaptive pathways' approaches: Scoping the literature. *Environmental Research Letters, 14*(9), 093007.

Butler, J. R. A., Suadnya, W., Yanuartati, Y., Meharg, S., Wise, R. M., Sutaryono, Y., & Duggan, K. (2016). Priming adaptation pathways through adaptive co-management: Design and evaluation for developing countries. *Climate Risk Management, 12*, 1–16.

Coulter, L. (2019). *User Guide for the climate change adaptation pathways framework: Supporting sustainable local food in B.C.* [User Guide]. https://www2.gov.bc.ca/assets/gov/farming-natural-resources-and-industry/agriculture-and-seafood/agricultural-land-and-environment/climate-action/user-guide-adaptation-pathways-preparedness.pdf

Ducrot, R. (2009). Gaming across scale in peri-urban water management: Contribution from two experiences in Bolivia and Brazil. *International Journal of Sustainable Development & World Ecology, 16*(4), 240–252.

Duke, R. (1980). A paradigm for game design. *Simulations and Games, 11*(3), 364–377.

Gomes, S. L. (2019). *An institutional approach to peri-urban water problems: Supporting community problem solving in the peri-urban Ganges Delta* [PhD dissertation, Delft

University of Technology]. https://repository.tudelft.nl/islandora/object/uuid%3A4e2900cd-1fa1-4bce-b0f5-c99f23a13c6c

Gomes, S. L. (2020). *An approach for participatory institutional analysis: A practical guide for capacity building with stakeholders* (p. 36) [Manual]. Delft University of Technology.

Gomes, S. L., & Hermans, L. M. (2018). Institutional function and urbanization in Bangladesh: How peri-urban communities respond to changing environments. *Land Use Policy, 79*, 932–941.

Gomes, S. L., Hermans, L. M., & Thissen, W. A. H. (2018). Extending community operational research to address institutional aspects of societal problems: Experiences from peri-urban Bangladesh. *European Journal of Operational Research, 268*(3), 904–917.

Greenblat, C. S. (1988). Setting objectives and parameters. In *Designing games and simulations: An illustrated handbook* (pp. 35–39). Sage Publications.

Haasnoot, M., Kwakkel, J. H., Walker, W. E., & ter Maat, J. (2013). Dynamic adaptive policy pathways: A method for crafting robust decisions for a deeply uncertain world. *Global Environmental Change, 23*(2), 485–498.

Hermans, L., & Cunningham, S. W. (2018). *Actor and strategy models: Practical applications and step-wise approaches*. Wiley & Sons.

Hossain, A. Z., Islam, K. F., Hossain, M. R., Hermans, L. M., & Sardjoe, N. (2019). *Capacity for participatory institutional analysis* (p. 29). Jagrata Juba Shangha.

Iaquinta, D. L., & Drescher, A. W. (2000). Defining periurban: Understanding rural-urban linkages and their connection to institutional contexts. *Tenth World Congress of the International Rural Sociology Association, 1*. http://www.ruaf.org/sites/default/files/econf1_submittedpapers_11iaquinta.doc

Johnson, M. P. (2012). Community-based operations Rresearch: Introduction, theory, and applications. In M. P. Johnson (Ed.), *Community-based operations research: Decision modeling for local impact and diverse populations* (pp. 3–36). Springer. https://doi.org/10.1007/978-1-4614-0806-2_1

Madani, K. (2010). Game theory and water resources. *Journal of Hydrology, 381*(3–4), 225–238.

Meijer, S. (2009). *The organization of transactions. Studying supply networks using gaming simulation* (Vol. 6). Wageningen Academic Publishers.

Narain, V. (2010). *Peri-urban water security in a context of urbanization and climate change: A review of concepts and relationships* (Discussion Paper No. 1; Peri Urban Project, pp. 1–16). http://saciwaters.org/periurban/idrc%20periurban%20report.pdf

North, D. C. (1990). *Institutions, institutional change and economic performance*. Cambridge University Press.

Ostrom, E. (2005). *Understanding institutional diversity*. Princeton University Press.

Rasmusen, E. (2007). The rules of the game. In *Games and information: An introduction to game theory* (pp. 11–33). Blackwell Publishing.

Roth, D., Khan, M. S. A., Jahan, I., Rahman, R., Narain, V., Singh, A. K., Priya, M., Sen, S., Shrestha, A., & Yakami, S. (2019). Climates of urbanization: Local experiences of water security, conflict and cooperation in peri-urban South-Asia. *Climate Policy, 19*(sup1), S78–S93.

Sen, S., Butsch, C., & Hermans, L. M. (2017). *H2O-T2S in urban fringe areas*. Project proposal for the "Transformations to Sustainability" programme. (p. 40). South Asia Consortium for Interdisciplinary Water Resources Studies (SaciWATERs).

Shaw, A. (2005). Peri-urban interface of Indian cities: Growth, governance and local initiatives. *Economic and Political Weekly, 40*(2), 129–136. Retrieved September 3, 2020, from http://www.jstor.org/stable/4416042

Slinger, J. H., Cunningham, S. W., Hermans, L. M., Linnane, S. M., & Palmer, C. G. (2014). A game-structuring approach applied to estuary management in South Africa. *EURO Journal on Decision Processes, 2*(3), 341–363.

United Nations. (2018). *World urbanization prospects: The 2018 revision—Key Facts* (Statistical Papers – United Nations) [Population and Vital Statistics Report]. United Nations. https://population.un.org/wup/Publications/Files/WUP2018-KeyFacts.pdf

Vervoort, J. M., Thornton, P. K., Kristjanson, P., Förch, W., Ericksen, P. J., Kok, K., Ingram, J. S. I., Herrero, M., Palazzo, A., Helfgott, A. E. S., Wilkinson, A., Havlík, P., Mason-D'Croz, D., & Jost, C. (2014). Challenges to scenario-guided adaptive action on food security under climate change. *Global Environmental Change, 28*, 383–394.

Chapter 9
Concluding Reflections: Towards Alternative Peri-Urban Futures?

Dik Roth and Vishal Narain

9.1 Introduction

This book represents the output of various research initiatives and projects that share a growing concern about urban expansion and the multiple ways in which it influences the city's surroundings, turning them into peri-urban and, ultimately, urban spaces. There are several good reasons for paying specific attention to the peri-urban. The past and current pace and trends of urbanization in South Asia — and more specifically the countries that feature in this book, Bangladesh, India and Nepal— foretell patterns of urban expansion that will deeply influence currently rural areas and populations. As an important dimension of this, the intensified exploitation of peri-urban natural resources like land, surface water and groundwater, and forests threatens the lives and livelihoods of peri-urban populations. As Swyngedouw and Kaika (2014, p.469) have argued, these "assemblages of capital-natures-cities-people", stretching far beyond the city itself, "retrace the socio-spatial choreographies of the flows of water, waste, food, etc., rearticulate patterns of control and access along class, gender and ethnic lines, and reconfigure maps of entitlement and exclusion". Despite all this, there is a strong tendency to approach these urbanization-related processes from an ecological modernization perspective, in which the city is a modern, sustainable and developmental win-win solution to the world's problems. Urban political ecology perspectives often show an urban focus as well, in which the metabolization of nature in urbanization processes is causally linked to the structural forces of capitalism and neoliberalism, but often not

D. Roth (✉)
Sociology of Development and Change group, Wageningen University,
Wageningen, The Netherlands
e-mail: dik.roth@wur.nl

V. Narain
Management Development Institute, Gurgaon, India
e-mail: vishalnarain@mdi.ac.in

researched from a peri-urban perspective. Hence, there is a need to pay attention to what these processes actually mean "on the ground" and how they are perceived and acted upon by people situated in specific peri-urban settings.

Within this general problematic of the peri-urban, our focus in this book was peri-urban water security, a topic that is receiving growing attention in South Asia and elsewhere. Important points of departure for researching water security are the need for an interdisciplinary approach rather than a disciplinary techno-managerial conceptualization of "scarcity", recognition of its situated, social and relational character, as well as the different experiences with and meanings given to water security. The dynamic peri-urban context requires research approaches to water security that take into account its emergence and fluidity, its ongoing production and reproduction in socio-natural transformations characterized by power differences, inequalities and in- and exclusions, creating multiple water securities and insecurities (see Boelens & Seemann, 2014; Lankford et al., 2013; Zeitoun et al., 2016).

In the remaining part of this chapter we reflect on the main findings from the various contributions to this book, all against the background of the wider debates and scientific insights on urbanization and water security. In the next section we summarize the main findings. Following this, we present a short reflection on peri-urban futures and the role of research and action in attaining them.

9.2 Summarizing the Main Findings

The chapters of this book are illustrations of various dimensions and manifestations of peri-urban water (in-)security, researched from various perspectives and representing different forms of engagement with different scientific and societal objectives in mind. They show the diversity and complexity of water security issues in the various case study locations, as well as their embeddedness in highly dynamic social, economic and other contexts and linkages with the wider urbanizing process. Multiple interrelated processes come together in the peri-urban: expansion of the resource needs and the growing ecological "footprint" of cities, resulting in intensified peri-urban resource exploitation and growing peri-urban environmental problems; an uprooting of existing livelihoods and lifestyles (from rural and agriculture-based towards urban and mainly non-agricultural), a process with both winners and losers; migration flows and other social-demographic changes; capitalist economic agendas and policies, with the state as a facilitator of global growth-based enterprise; and, last but not least, a changing climate.

The examples of these insecurities are manifold. Flows of water into cities to meet growing urban water demands are changing peri-urban water rights, water control and access, as well as water availability (e.g. the drinking water canals discussed in Chap. 6, and commercial water provision through tankers and groundwater in Chaps. 5 and 8). Return flows of wastewater or disposal of solid waste in peri-urban water bodies have become a major environmental threat to peri-urban water security (see its linkages with lakes and wetlands in Chap. 2, and its role in conflicts around gate operation in Chap. 7). At the same time, they are sometimes seen —and used— as an opportunity for adaptation of peri-urban agriculture and aquaculture to changing peri-urban conditions (e.g. Chaps. 2 and 8) and to a changing climate (Chap. 6). Massive migration into peri-urban spaces, the emergence of new markets for agricultural produce and new water needs change locally existing water rights, forms of management, functions and uses of water (see the case of surface irrigation in Kathmandu Valley; Chap. 3). In the socio-economic mix of global capitalism and local hierarchy and patriarchy of the ready-made garment industry, the challenges experienced by women in securing a living place with secure access to domestic water are huge, in sharp contrast to optimistic assumptions of their "empowerment" (Chap. 4).

Above all, the chapters of this book show how peri-urban water users are continuously and creatively engaging with the multiple water-related challenges emerging in their life-worlds. The inhabitants of peri-urban spaces are far from passive onlookers, but actively try to come to terms with changing water security, devise solutions and look for opportunities (see Long, 2001). They adapt their agricultural, aquacultural and other water-based livelihoods, explore the opportunities provided by new infrastructure (such as wastewater canals) and investments in new water technology (e.g. groundwater pumping devices; drip and sprinkler irrigation; wastewater appropriation and pumping devices). They try to institutionalize new practices of water use around canals, and adapt their cropping choices and schedules to changing water security. They organize in new ways to seek forms of government or project support for investments, awareness raising and capacity building. They explore new urban markets for agricultural produce (and for products like bottled water or bricks, for that matter; all activities that may further increase water scarcity and competition). They engage in conflict negotiation and resolution, and adapt their water use practices to avoid conflicts in conflict-prone situations of water competition. The mediation of peri-urban water insecurity is a socio-technical process, in which both technologies and institutions are mobilized (see Roth & Vincent, 2013).

Such adaptive solutions, however, are not accessible for all; win-wins hardly exist in real life. As several contributions have shown, urbanization creates opportunities for some, but also reproduces or worsens existing forms of social differentiation, inequalities and related water insecurities. It crucially changes existing water control, rights and access, and does so in unequal ways, in the process creating multiple in- and exclusions (Bartels et al., 2020). In Kathmandu Valley (Chap. 3), for instance, dependence on an increasingly unreliable canal irrigation system can be reduced by using alternative technologies, such as groundwater pumps. Such solutions, however, are only available to those who can afford the investments in this technology (or other, water-saving, alternatives). Massive pumping may in the long run reduce groundwater levels and threaten sustainability of groundwater use, creating new inequalities between those who can invest in more pumping power and those who cannot (see also Shrestha, 2019a).

Institutionally, the peri-urban is a hybrid space characterized by problems of existence of policy gaps, overlaps, ambivalences, contradictions and conflicts between state-initiated institutional arrangements that are either "urban" or "rural", and local ones that are embedded in specific situated ways of governing, managing and using resources, defining rights and restrictions, jurisdictions and authorities. The resulting competing frameworks, contradictions and gaps in peri-urban governance, combined with the intensified exploitation of resources by a variety of actors with often competing interests, make the exploitation of important resources in these spaces conflict-prone. This legal-institutional complexity makes for a property landscape characterized by "fuzziness" of property relations, allowing for speculative claiming of access and rights to resources, either actively supported or silently condoned by government administrations and agencies that refrain from actively intervening. Several chapters noted the problem of the disappearing or degrading commons under various processes of encroachment and privatization, and its consequences for people depending on them (see also Narain & Vij, 2016).

Developments in the peri-urban space certainly increase the risk of resource-related conflicts. On the basis of the research reported in this book, however, there is no evidence to support simplistic assumptions about a direct scarcity - conflict causality. The chapters that explicitly dealt with issues of conflict and cooperation are, of course, far from conclusive. Though resource-related conflicts occur regularly, they seem to seldom turn violent and cannot be causally related to an abstract notion of growing "scarcity". Some observations seem to be possible. In several cases discussed, the relatively gradual processes of change for which it may be difficult to directly allocate responsibility locally may be part of the explanation for the fact that people tend to seek adaptive options rather than engaging in conflicts. In more sensitive cases, where the effects of people's actions are directly felt, conflict avoidance may be preferred in order to maintain good relationships and decrease dependence on a contested water source. In the Nepal case (Chap. 3), for instance, human-induced water scarcity does not lead to conflict but to more investments in technology, dug wells and borewells. Hence more intensive water use does not necessarily lead to more conflict. In the case of gate operation in Bangladesh (Chap. 7), the authors note that people are losing interest in negotiation and dialogue about the gate operation because the river water quality has degraded to such an extent that it is rapidly becoming useless.

As for cooperation, several cases show the important role of locally existing norms, rule systems and practices in defending existing water rights (even if with limited success only, such as in Nepal) and in establishing and strengthening new forms of collective action, including defining rights and responsibilities (the case of wastewater canals in India). Last but not least — and in contrast with the simplistic conflict-cooperation dichotomy, which does not grasp the multiple dimensions, layers and nuances of both — much cooperation is invisible but exists in the networks of interest and power that normalize existing relations of power and perpetuate forms of dependence, exploitation, and inequalities. The research presented in this book suggests that conflicts and co-operation do not exist as binaries; rather there exists a continuum representing varying degrees and forms of co-operation and conflict. Other than conflict or co-operation, there could be situations of conflicts of interest or forced co-operation (see also Vij et al., 2018). Power relations may prevent the escalation of conflicts of interest into explicit conflicts. The dependence of sharecroppers on landowners and of water users on providers of water, for instance, are relationships and mechanisms that dampen conflict.

9.3 Peri-Urban Futures: From Local Struggles to Transformative Changes?

9.3.1 Engaging with the Peri-Urban: Rearguard Action?

Several chapters show specific intervention-based concerns with the peri-urban and the management and protection of its resources for the benefit of peri-urban populations. Are such concerns for peri-urban land and water rights, irrigation canals, commons and livelihoods more than fighting a rearguard struggle, just before the bulldozers and concrete mixers definitively roll out the city? Take, for instance, local engagement with the preservation of irrigation canals in Kathmandu Valley (Chap. 3). It is highly probable that such engagement will not save this canal or others, nor the related livelihoods and agricultural practices, from the waves of urban expansion that roll on and face few policy restrictions. Just a few more years, and these peri-urban spaces will probably have become fully urbanized. In a similar way, the authors of Chap. 2 (Mundoli et al.) plead for a change in perception with regard to waste-water linkages between cities and their peri-urban spaces and water. But to what extent does this contribute to the implementation of an ecological modernization-style urban sustainability agenda that comes close to what Kaika (2017, p.98) has called "immunological practices" — combating the symptoms but not the deeper causes of the problem?

Every form of engagement raises questions as to which perspectives, approaches and activities stand a chance of contributing to socially, politically and environmentally sustainable peri-urban transformations. What is realistically possible and worth struggling for, and how? How can "the right to the peri-urban" be defended in a socially, politically and environmentally meaningful way? Should such struggles aim for quickly achievable small improvements in local issues, or link up to bigger concerns, networks and movements? How can local peri-urban engagement and initiatives be scaled up and connect to broader initiatives or social movements on socio-environmental and political issues of urbanization? In short: how can peri-urban rearguard action become part of a truly transformative social-environmental political movement that transcends the artificial boundaries of the urban, the peri-urban and the rural? We do not pretend to have the answers to these question, but will shortly reflect on a number of research and action needs for the peri-urban.

9.3.2 Research Needs for the Peri-Urban Space

What kind of alternative peri-urban futures are realistically imaginable, and what kinds of scientific engagement are needed to take steps in realizing them? There is a growing scientific, NGO and broader societal engagement with peri-urban issues in all three countries discussed in this book. Governments are increasingly made aware of the specific characteristics and problems of the peri-urban, and in some cases show growing recognition of the need to engage with these problems. Overall, however, governmental engagement (or rather non-engagement) with the peri-urban seems to be part of the problem rather than of the solution. As discussed above, we should not have unrealistic expectations about the role of new administrative divisions, laws and policies. Yet a key step would be a political-administrative recognition of the specifically peri-urban dimensions of urbanization, their influences on peri-urban livelihoods and natural resources, and the need for governance approaches that take these peri-urban characteristics as a point of departure for rethinking the urban-rural dichotomies that form the usual basis of administrative divisions, policies and forms of legal regulation. Making the peri-urban visible as a fluid, hybrid and institutionally complex space with specific problems that need policy attention remains crucial (see Allen, 2003; Allen et al., 2006). There is a clear role for scientific engagement and active science-policy interaction here. It is important, however, that scientists and researchers remain critical of the ways in which relevant and potentially transformative scientific developments, insights and concepts are turned into policy buzzwords, instruments and objectives.

An example of this is the relationship between scientific research and the changing agendas and priorities of development policies. Several authors (e.g. Arabindoo, 2009; Kaika, 2017) have noted the important role of development funding and global development institutions in setting agendas and priorities for research and policy on themes like urbanization, peri-urban research and climate-related research. Kaika (2017) mentions the focus on "safe, resilient, sustainable and inclusive cities" that has become part of the current urban Sustainable Development Goals agenda. Like "sustainability", "resilience" has become a popular buzzword in the development policy world, as part of what Taylor (2014) has called "the holy trinity" of climate change adaptation. Although this is not the place to go into the criticism of the superficial and depoliticizing uses of concepts like resilience (see e.g. Béné et al., 2014; Boyden & Cooper, 2007; Taylor, 2014), important lessons can be learned from the ways in which such concepts are taken up and are given meaning (and power) as they "travel" through the development policy world. Much more critical thinking about such developmental trends and fashions is needed on the part of research funders, scientific institutions and researchers, to put into perspective its uses and claims of its relevance, and criticize its depoliticizing effects on debates about sustainability, climate change and urbanization, including the peri-urban (see Shrestha, 2019a, b).

Urban political ecology has yielded extremely important insights on urbanization and the socially unequal metabolic flows that interconnect multiple scales and spaces far beyond cities themselves. It has also criticized the techno-managerial and post-political character of market-led sustainability approaches (Cook & Swyngedouw, 2012; Swyngedouw & Heynen, 2003). However, as discussed above, even in this literature there is an urban bias (Angelo & Wachsmuth, 2014, 2020), while local peri-urban manifestations, processes and mechanisms of urban – peri-urban metabolic flows, researched as "assemblages of capital-natures-cities-people" (Swyngedouw & Kaika, 2014) remain a black box. Future research can fill a gap here by becoming explicitly ethnographic, studying in an in-depth way the multiple dimensions of peri-urban social-environmental changes as situated in specific socio-economic, cultural and other contexts in which these changes are experienced and acted upon in multiple and unequal ways (see Shrestha, 2019a; Webster, 2011).

However, in light of the many insights from urban political ecology on the scalar relationships between urbanization and the metabolic processes in the peri-urban space, research should also move beyond the local and, where research links up with action, beyond consensus-based institutional design approaches. It should trace the linkages and flows of urbanization by "studying up", crossing spatial and other boundaries and actively engaging with the urban-based actors, powers and policies that propagate specific forms of urban expansion, economic growth and resource exploitation beyond the city, in short: the political-economic processes that fuel the engine of urban expansion and lead to contestations between state and corporate power on one hand, and differently affected communities on the other (Shatkin, 2019). Instead of dealing with such contestations through the usual consensus-building approaches (e.g. stakeholder platforms, participation, co-creation), Kaika (2017, p.99) proposes to take issues of dissensus as "living indicators" and "signposts" for further action research and political engagement.

Last but not least, research should put much more effort at in-depth research on resource (re-)allocations, property transfers and property transformations that are so deeply influencing the peri-urban space. Degradation and disappearance of the commons, for instance, is frequently mentioned as an important impact of urbanization, but in-depth studies on the processes and mechanisms through which such transformations can take place are scarce. Aside from the fact that such transformations are not unique to the peri-urban space, they are also rooted in wider socio-environmental transformations that may throw light on causes of their degradation and disappearance. The commons has become an ideological concept with multiple meanings, but what is "the commons" in specific socio-historical and resource use contexts? What processes of claiming and counter-claiming are developing around commons? How are property relations defined and redefined, and turned into use and management practices? How are commons given new meaning? What has disappeared, and why? What do we want the commons to be(come)? It is only through such understandings of the commons that we can start using the commons concept as a social imaginery (Wagner, 2012; see Bakker, 2010) and envisioning alternative peri-urban commons futures.

9.3.3 Research, Policy and Action

As brought out by all contributions in this volume, the peri-urban is a complex space to intervene. Approaches to intervention are based on the narrative that there is a need to straddle the rural-urban dichotomy in development see also Mehta and Karpouzoglou (2015). This needs overcoming the divides created between urban planning and rural development agencies among the institutions of the state. While this is important given the fluid, transitory nature of peri-urban contexts, it is simplistic to rely only on formal state approaches to address peri-urban challenges. As some contributions in this volume show, interventions in the peri-urban space need strong coalitions among critical academics, civil society organizations (CSOs), citizens and state agencies, combining research with action. CSOs active in the field have a strong grounding in local contexts and an ability to engage with and mobilise local populations. Academics and researchers possess the skills for scientific research and documentation. This academic-CSO nexus is necessary to reorient state agencies, create mutual accountability relationships between peri-urban populations and state agencies, and sensitize the state to seriously addressing peri-urban issues. At the same time, CSOs must continue to play a role in providing support to peri-urban populations, getting them into dialogue with state agencies, building their capacity to demand better and improved services, and helping them to defend their rights where needed.

While the literature on gender and water has grown well in recent decades, little is known about changing gender relations around water in peri-urban spaces. This constitutes another area for further research and action. Given the social heterogeneity in peri-urban contexts, intersectionality of gender with other axes of social differentiation can be very sharp indeed and merits further investigation, in relation to agendas for action to deal with such forms of social differentiation. Given the gender-based division of labor around collection of water and other water-related tasks, this needs specific attention in action-research concerning the water-related transformations of the peri-urban and changing water security.

Another important domain in which research and action should combine is in uncovering the vulnerabilities that are created or reproduced by urbanization processes. These could become the basis for sustained advocacy, political action and policy reforms, in which linkages are sought with, for instance, water justice and environmental justice movements that are active at higher levels and may connect urban and peri-urban issues. Although peri-urban water security is always situated and contextual, we see a need here for initiatives to protect the peri-urban to engage with supra-local movements and thus to create forms of engagement that transcend the boundaries of localised "particularisms" (Harvey, 1996; see Walker, 2009). Such movements could also play a role in linking problems of peri-urban water security and public water provision to (hydraulic) citizenship (see Anand, 2011; Gandy, 2004). Finally, such a broader movement linking peri-urban and urban issues can contribute to exposing the fallacies and downsides of the techno-managerial "smart cities" ideology.

References

Allen, A. (2003). Environmental planning and management of the peri-urban interface: Perspectives on an emerging field. *Environment and Urbanization, 15*(1), 135–147.

Allen, A., Dávila, J. D., & Hofmann, P. (2006). The peri-urban water poor: Citizens or consumers? *Environment and Urbanization, 18*(2), 333–351.

Anand, N. (2011). Pressure: The politechnics of water supply in Mumbai. *Cultural Anthropology, 26*(4), 542–564.

Angelo, H., & Wachsmuth, D. (2014). Urbanizing urban political ecology: A critique of methodological cityism. *International Journal of Urban and Regional Research, 39*(1), 16–27.

Angelo, H., & Wachsmuth, D. (2020). Why does everyone think cities can save the planet? *Urban Studies, 57*(11), 2201–2221.

Arabindoo, P. (2009). Falling apart at the margins? Neighbourhood transformations in peri-urban Chennai. *Development and Change, 40*(5), 879–901.

Bakker, K. (2010). *Privatizing water. Governance failure and the world's urban water crisis.* Cornell University Press.

Bartels, L. E., Bruns, A., & Simon, D. (2020). Towards situated analyses of uneven peri-urbanisation: An (urban) political ecology perspective. *Antipode, 52*(5), 1237–1258.

Béné, C., Newsham, A., Davies, M., Ulrichs, M., & Godfrey-Wood, R. (2014). Review article: Resilience, poverty and development. *Journal of International Development, 26*(5), 598–623.

Boelens, R. A., & Seemann, M. (2014). Forced engagements: Water security and local rights formalization in Yanque, Colca Valley, Peru. *Human Organization, 73*(1), 1–12.

Boyden, J. & Cooper, E. (2007). *Questioning the power of resilience: Are children up to the task of disrupting the transmission of poverty?* Chronic Poverty Research Centre (CPRC) working paper 73. University of Oxford.

Cook, I. R., & Swyngedouw, E. (2012). Cities, social cohesion and the environment: Towards a future research agenda. *Urban Studies, 49*(9), 1959–1979.

Gandy, M. (2004). Rethinking urban metabolism: Water, space and the modern city. *City, 8*(3), 363–379.

Harvey, D. (1996). *Justice, nature & the geography of difference.* Blackwell.

Kaika, M. (2017). 'Don't call me resilient again!': The new urban agenda as immunology … or … what happens when communities refuse to be vaccinated with 'smart cities' and indicators. *Environment and Urbanization, 29*(1), 89–102.

Lankford, B., Bakker, K., Zeitoun, M., & Conway, D. (Eds.). (2013). *Water security: Principles, perspectives, practices.* Routledge/Earthscan.

Long, N. (2001). *Development sociology: Actor perspectives.* Routledge.

Mehta, L., & Karpouzoglou, T. (2015). Limits of policy and planning in peri-urban waterscapes: The case of Ghaziabad, Delhi, India. *Habitat International, 48*, 159–168.

Narain, V., & Vij, S. (2016). Where have all the commons gone? *Geoforum, 68*, 21–24.

Roth, D., & Vincent, L. (Eds.). (2013). *Controlling the water. Matching technology and institutions in irrigation management in India and Nepal.* Oxford University Press.

Shatkin, G. (2019). The planning of Asia's mega-conurbations: Contradiction and contestation in extended urbanization. *International Planning Studies, 24*(1), 68–80.

Shrestha, A. (2019a). *Urbanizing flows. Growing water insecurity in peri-urban Kathmandu Valley, Nepal.* PhD thesis, Wageningen University, Wageningen, the Netherlands.

Shrestha, A. (2019b). Which community, whose resilience? Critical reflections on community resilience in peri-urban Kathmandu Valley. *Critical Asian Studies, 51*(4), 493–514.

Swyngedouw, E., & Heynen, N. C. (2003). Urban political ecology; justice and the politics of scale. *Antipode, 35*(5), 898–918.

Swyngedouw, E., & Kaika, M. (2014). Urban political ecology. Great promises, deadlock… and new beginnings? *Documents d'Anàlisi Geogràfica, 60*(3), 459–481.

Taylor, M. (2014). *The political ecology of climate change adaptation: Livelihoods, agrarian change and the conflicts of development.* Routledge / Earthscan.

Vij, S., Narain, V., Karpouzoglou, T., & Mishra, P. (2018). From the core to the periphery: Conflicts and cooperation over land and water in periurban Gurgaon, India. *Land Use Policy, 76*, 382–390.

Wagner, J. R. (2012). Water and the commons imaginary. *Current Anthropology, 53*(5), 617–641.

Walker, G. (2009). Beyond distribution and proximity: Exploring the multiple spatialities of environmental justice. *Antipode, 41*(4), 614–636.

Webster, D. (2011). An overdue agenda: Systematizing East Asian Peri-urban research. *Pacific Affairs, 84*(4), 631–642.

Zeitoun, M., Lankford, B., Krueger, T., Forsyth, T., Carter, R., Hoekstra, A., Taylor, R., Varis, O., Cleaver, F., Boelens, R., Swatuk, L., Tickner, D., Scott, C., Mirumachi, N., & Matthews, N. (2016). Reductionist and integrative research approaches to complex water security policy challenges. *Global Environmental Change, 39*, 143–154.